Materials and Physics
for Nonvolatile Memories

MATERIALS RESEARCH SOCIETY
SYMPOSIUM PROCEEDINGS VOLUME 1160

Materials and Physics for Nonvolatile Memories

Symposium held April 14–17, 2009, San Francisco, California, U.S.A.

EDITORS:

Yoshihisa Fujisaki

Hitachi Ltd.
Kokubunji, Tokyo, Japan

Rainer Waser

RWTH Aachen University
Aachen, Germany

Tingkai Li

Sharp Laboratories of America
Camas, Washington, U.S.A.

Caroline Bonafos

CEMES/CNRS
Toulouse, France

Materials Research Society
Warrendale, Pennsylvania

CAMBRIDGE UNIVERSITY PRESS
Cambridge, New York, Melbourne, Madrid, Cape Town,
Singapore, São Paulo, Delhi, Mexico City

Cambridge University Press
32 Avenue of the Americas, New York NY 10013-2473, USA

Published in the United States of America by Cambridge University Press, New York

www.cambridge.org
Information on this title: www.cambridge.org/9781107408296

Materials Research Society
506 Keystone Drive, Warrendale, PA 15086
http://www.mrs.org

First published 2009
First paperback edition 2012

Single article reprints from this publication are available through
University Microfilms Inc., 300 North Zeeb Road, Ann Arbor, MI 48106

CODEN: MRSPDH

ISBN 978-1-107-40829-6 Paperback

*Invited Paper

RESISTIVE SWITCHING RAM I

RESISTIVE SWITCHING RAM II

POSTER SESSION:
RESISTIVE SWITCHING RAM III

*Invited Paper

PREFACE

Symposium H, "Materials and Physics for Nonvolatile Memories," held April 14–17 at the 2009 MRS Spring Meeting in San Francisco, California, was the third in a series of MRS symposia on nonvolatile memories. Progress in the technical development of many kinds of nonvolatile memories since the previous symposium in 2007 was highlighted during this symposium.

Research results on Advanced Flash Memories including nano-particle floating gate FETs, MRAM, FeRAM, ReRAM and Phase Change RAMs as well as memories using polymer materials were presented. The papers in this volume are published in the order they were presented during the symposium. The advanced Flash memory and charge trap memory chapters include the state of the art in the post Flash memory technologies. The charge trap memory chapter shows the challenges to make a brake though the limitation of high density Flash memories. The field of resistance switching materials is maturing rapidly as demonstrated by the excellent overview and summary of ReRAM switching mechanisms in the Resistive Switching RAM chapter. The Ferroelectric Nonvolatile Memory Devices chapter focuses on the advances in materials and device demonstrations for both capacitor type and ferroelectric gate memory approaches to achieve high density ferroelectric memories. The Phase Change Nonvolatile Memory Devices chapter describes a wide variety of materials that continue to be considered for phase change and resistance switching memories, while the understanding of the most established materials are quickly maturing. The Magnetic Resistive RAM chapter includes papers from a joint session between this symposium and Symposium FF, "Novel Materials and Devices for Spintronics."

The strong and increasing interest in nonvolatile memories, both domestic and international, indicates the worldwide importance of these materials and memory devices. This symposium proceedings volume represents the latest technical advances and information on nonvolatile memory devices from universities, national laboratories and industry. It also provides insight into emerging trends in these exciting technologies.

We would like to thank all of the speakers and participants for their valuable contributions toward making the symposium successful. We gratefully acknowledge the financial support of ANEALYSYS, Ion Beam Services, Park Systems, SVT Associates, Inc. and Universal Systems.

Yoshihisa Fujisaki
Rainer Waser
Tingkai Li
Caroline Bonafos

August 2009

MATERIALS RESEARCH SOCIETY SYMPOSIUM PROCEEDINGS

MATERIALS RESEARCH SOCIETY SYMPOSIUM PROCEEDINGS

Advanced Flash I

Mater. Res. Soc. Symp. Proc. Vol. 1160 © 2009 Materials Research Society   1160-H01-03

Ultra-Low Energy Ion Implantation of Si Into HfO2-Based Layers for Non Volatile Memory Applications

P. E. Coulon[1], K. Chan Shin Yu1*, S. Schamm[1], G. Ben Assayag[1], B. Pecassou[1], A. Slaoui[2], S. Bhabani[2], M. Carrada[2], S. Lhostis[3] and C. Bonafos[1]

[1]Groupe Nanomat – CEMES-CNRS – 29 rue J. Marvig - BP 94347 - 31055 Toulouse Cedex 4 – Université de Toulouse
[2]InESS – CNRS, 23 rue du Loess, 67037 Strasbourg, France
[3] ST Microelectronics, 850 rue Jean Monnet, 38926 Crolles
* now at LAAS-CNRS, Université de Toulouse

ABSTRACT

The fabrication of NCs is carried out using an innovative method, ultra-low energy ($\leq$5 keV) ion implantation (ULE-II) into thin (6-9 nm) HfO2–based layers in order to form after subsequent annealing a controlled 2D array of Si NCs. The implantation of Si into HfO2 leads to the formation of SiO2–rich regions at the projected range due to the oxidation of the implanted Si atoms. This anomalous oxidation that takes place at room temperature is mainly due to humidity penetration in damaged layers. Different solutions are investigated here in order to avoid this oxidation process and stabilize the Si-phase. Finally, unexpected structures as HfO2 NCs embedded with SiO2 matrix are obtained and show interesting memory characteristics. Interestingly, a large memory window of 1.18 V has been achieved at relatively low sweeping voltage of ± 6 V for these samples, indicating their utility for low operating voltage memory device.

INTRODUCTION

Nanocrystal memory (NCM) devices are competitive candidates for extending further the scalability of Flash-type memories [1-3]. Various process/materials alternatives have been suggested recently to establish a proven NCM technology in the timeframe required by the industry roadmap. In this direction, the fabrication of NCs into high-k dielectric matrices instead of SiO2 materials has retained particular attention for achieving NCMs with low programming voltages and improved data retention. Indeed, a dielectric with a higher dielectric constant allows, in principle, to use a thicker tunnel oxide reducing leakage currents and, because of the smaller conduction band offset between the high-k film and the silicon substrate, to achieve the goal of a low voltage non-volatile memory device.

Among the different high-k materials under investigation, HfO2 and their alloys are probably the most widely studied and are considered as very promising candidates for the integration in ultra-scaled commercial devices. Most of the data reported in literature on the synthesis of metallic and semiconducting nanocrystals for memory device application focus on the possibility to integrate the nanocrystals in these materials. Very few data are available on the synthesis of Si nanocrystals in a high-k matrix, especially in Hafnium oxide alloys. Superior programming efficiency and data retention characteristics have been obtained for systems described by Si nanocrystals deposited on HfO2 films by LPCVD with SiH4 at 600°C followed by in-situ HfO2 deposition of the control dielectric [4-5]. Promising device results using Si or Ge NCs embedded in HfO2 or HfAlO gate dielectrics have been recently presented [6-7]. Nevertheless, the fabrication of semiconducting NCs in high-k materials remains not straightforward. The main problems in terms of material science related to the use of a matrix other than SiO_2 for the fabrication of memories are identified in ref. [8], and mainly concern the diffusion and oxidation of Si in high-k (HfO_2 and Al_2O_3) matrix.

Recently, we have extensively used ultra-low energy ion implantation (ULE-II) to synthesize single planes of Si nanocrystals embedded in very thin (5 to 10 nm) oxide layers. The depth-location of these two-dimensional (2D) arrays of particles below the wafer surface can be controlled with nanometer precision by tuning the implantation energy while their size and density can be controlled by varying the dose and annealing conditions [9]. These parameters were finally optimized to fabricate non volatile memory devices of improved characteristics [3]. In this paper, we extend this approach to synthesize a plane of Si NCs within high-k dielectrics. In particular, we present an approach to face the challenge of Si NCs formation into very thin (5-10 nm) HfO_2-based layers. This approach relates to the fabrication by ultra-low energy ion-beam-synthesis (ULE-IBS) of Si NCs in HfO_2 and HfSiO layers deposited by MOCVD on Si wafers. After implantation, all samples are annealed at high temperature for the purpose of NCs formation. Structural and chemical studies are carried out at the atomic scale by High Resolution Transmission Electron Microscopy (HREM) and Electron Energy Loss Spectroscopy in scanning mode by using a nanometer probe (STEM-EELS). The structural study reveals the oxidation of most of the implanted Si in the HfO_2-based layers while Si NCs are formed in the interfacial SiO_2 layer. Different strategies are tested in order to limit this oxidation including NH_3 annealing, implantation through thin Si_3N_4 capping layers and N and Si co-implantation in HfO_2 based layers. Unexpected structures as HfO_2 NCs embedded with SiO_2 matrix are obtained in addition to Si NCs in the SiO_2 interfacial layer (IL). In this case, capacitance-to-voltage characteristics of the MIS capacitors with NCs revealed strong hysteresis in terms of flat-band voltage shift after application of gate-voltage round-sweeps. These results suggest charge trapping and storage related to the formation of the NCs through the implanted/annealed high-k layers.

EXPERIMENT

As-deposited HfO_2 and HfSiO layers have been grown by MOCVD and are both nanocrystalline with very small grains of 2 nm. They are respectively 7 and 6 nm thick and are separated from the substrate by a very thin SiO_2 IL of 1 nm thick (not shown). Si+ has been implanted at low energy and high doses in these layers. The implantation energy has been chosen by running TRIM simulations [10] such as the projected range of the implanted Si is located in the middle of the high-k layer. The dose is higher than the nucleation threshold for the system Si/SiO_2 corresponding to 10 at.% [3] but lower than the threshold value for strong interface mixing effects [9]. For each layer, the implantation conditions are summarized in Table 1. On some HfO_2 layers, an additional 5 nm-thick layer of Si_3N_4 was deposited by electron cyclotron resonance-chemical vapor deposition (ECR-CVD) before implantation in order to prevent the Si oxidation. The samples have been further annealed at 1000°C for 30s under N_2 ambient for the purpose of Si-NCs formation except one which has been annealed under NH_3 at 950°C for 10 minutes.

The resulting structures were examined by High Resolution Transmission Electron Microscopy (HRTEM) in cross-section (XS) preparation to evidence the phase separation, as well as to determine the crystallographic nature of the NCs and the degree of crystallization of the surrounding matrix. Chemical images are performed by Energy Filtered TEM (EFTEM) and Electron Energy Loss Spectroscopy (EELS) with a probe of 1 nm used in scanning mode (STEM-EELS) to analyze the composition of the different layers. For this study, the cross-section samples were prepared for TEM observations using the standard procedure involving mechanical polishing and ion milling. Images were taken using a field emission TEM, FEI Tecnai™ F20 microscope operating at 200 kV, equipped with a spherical aberration corrector and the last generation of the Gatan imaging filter series, the TRIDIEM. The spherical aberration corrector allows high quality HREM images with an increased signal/noise ratio and nearly no delocalization effect.

Prior to electrical measurement, 100 nm Al gate electrodes having a diameter of ~ 0.75 mm were thermally evaporated into the samples through a shadow mask. For ohmic contact, gold was evaporated on the back side of Si substrates. High frequency capacitance-voltage (C-V) measurements of metal-insulator-semiconductor (MIS) structures were carried out using a HP4192A impedance analyzer. Post-metallization annealing of all the MIS structures were carried out at 400°C for 20 minutes in forming gas prior to measurement. Control MIS structures of the gate dielectrics without undergoing the implantation step were also fabricated using the same processes as the implanted samples and examined for comparison purposes.

film	thickness (nm)	grain size (nm)	implant	energy (keV)	dose ($\times 10^{16}$ at.cm^{-2})	annealing conditions
HfO$_2$	7	2 nm	Si	3	1.5	1000°C 30s N$_2$
HfO$_2$	9	4 nm	Si	5	1.5	1000°C 30s N$_2$
HfO$_2$	7	2 nm	N+N+Si	1+2.5+3	0.3+0.7+1.5	1000°C 30s N$_2$
HfO$_2$	9	4 nm	N+N+Si	1+3+5	0.2+1.0+1.5	1000°C 30s N$_2$
HfO$_2$+Si$_3$N$_4$	7+5	2 nm	Si	8	1.5	1000°C 30s N$_2$
HfO$_2$+Si$_3$N$_4$	9+5	4 nm	Si	9	2	1000°C 30s N$_2$
HfO$_2$	7	2 nm	Si	3	1.5	950°C 10min NH$_3$
HfO$_2$	9	4 nm	Si	5	1.5	950°C 10min NH$_3$
HfSiO	6	2 nm	Si	2	1	1000°C 30s N$_2$

Table 1: implantation and annealing conditions of the films discussed in this work.

RESULTS AND DISCUSSION

Fig. 1(a) shows a Bright Field XS-TEM image of the HfO$_2$ layer after Si+ implantation and subsequent annealing. The SiO$_2$ interfacial layer (IL) separating the HfO$_2$ layer and the Si substrate is 7.2 nm instead of 1.2 nm for the as-deposited layer. The HfO$_2$ layer, in dark contrast, is now polycrystalline (see HREM image in Fig. 1b) with nanograins of 4 nm in diameter. The high-k layer is 10.9 nm thick i. e., has encompassed a swelling of 4 nm comparing to the as-deposited layer. Regions with clear contrast, very similar to the one in the IL, are visible in the high-k layer. They are amorphous (see Fig. 1b), Si-rich and separate the HfO$_2$ grains.

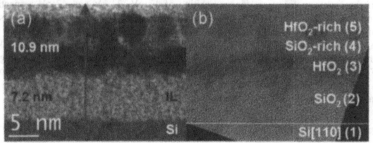

Fig. 1: (a) BF and (b) HREM images of the HfO$_2$ layer implanted with Si+ at 3 keV for a Si excess of 26 at. %. The HfO$_2$ grains are separated by amorphous Si rich regions. The red arrow corresponds to the line scanned across the film for the STEM-EELS acquisition.

STEM-EELS analyses have been performed in order to examine the different EELS signatures in the low-energy loss domain and around the core Si-K edge in order to identify the corresponding phases. The experiment consists in scanning a small probe 1 nm in diameter across the interface with the Si substrate going towards the film. An EELS spectrum is recorded at each position of the probe along the line. The analysis in the core-loss region (Hf-M and Si-K) shows that both the IL and the clear regions separating the HfO_2 grains are SiO_2-rich (Fig. 2a) while from the low-loss analysis, beyond the signatures of the Si substrate, the SiO_2 IL and the HfO_2-based nanocrystals, a signature characteristic of the plasmon of Si has been detected in addition to the one of the SiO_2 IL layer (Fig. 2b).

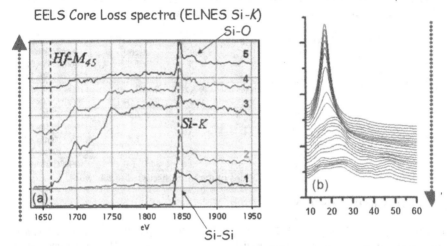

Fig. 2: STEM-EELS analyses (line-spectrum) of the different regions as indicated by the arrow on Fig. 1(a), (a) in the core-loss region and (b) in the low-loss region. The number in (a) correspond to the area defined in Fig. 1b.

In order to localize within the film where are located the different phases identified by STEM-EELS, EFTEM chemical images have been performed. The images are formed with the electrons that are selected by a slit with a width of +/-2 eV placed in the energy-dispersive plane of the spectrometer at an energy position of 17 eV, 23 eV and 47 eV respectively, corresponding to the Si and SiO_2 plasmon peak positions for the formers and a strong contribution of the Hf-$O23$ edge in HfO_2 for the last one. Isolated grains of HfO_2 with white contrast are revealed with 47 eV energy filtering (Fig. 3c). The SiO_2-rich (SiO_x) areas like the IL and regions between the HfO_2 grains can be recognized by 23 eV energy filtering (Fig. 3b). The EFTEM at 17 eV is very interesting, showing no visible contrast in the HfO_2 region but Si NPs in the IL (Fig. 3a). These NPs, with a diameter of 1-2 nm, are located in a 3 nm-thick region at 1.5 nm from the Si substrate and 3 nm from the first HfO_2 NCs.

Fig. 3: EFTEM images obtained by filtering around (a) the Si Plasmon (17 eV), (b) SiO_2 plasmon (23 eV) and (c) strong contribution of Hf-*O23* (47 eV).

As a consequence, one can conclude that the majority of the Si implanted in the HfO_2 region has been oxidised, giving rise to the separation of the HfO_2 grains. The swelling measured for the high-k layer after implantation and annealing is 4 nm, i.e. larger than the 3 nm expected for only matter addition of the total dose. This is in good agreement with the partial oxidation of the implanted Si, the dramatic swelling of the IL (expansion of 6 nm after implantation and annealing) being then due to the oxidation of the Si substrate. This oxidation of the Si implanted in the HfO_2 layer already takes place before annealing (not shown). During annealing, part of the implanted Si diffuses in the SiO_2-IL and precipitates to form Si NCs.
The oxidation of implanted Si has been studied in detail when implanted in very thin SiO_2 matrix [10-13]. This so-called "anomalous" diffusion takes place before any annealing, immediately after implantation because the heavily damaged SiO_2 layers absorb humidity. These water molecules are driven in the layers, dissociate and finally react with the implanted Si to form SiO_2. During annealing, further dissociation of OH takes place and finally most of the H atoms diffuse out to the surface. The penetration and final concentration of water molecules do not depend on the relative humidity in the atmosphere but are only limited by the degree of damage i. e., by the concentration of defects, in the SiO_2 matrix. In the case of Si implanted in SiO_2, we have evaluated that 40% of the implanted Si is oxidised. The remaining Si precipitates to form Si NCs. In the case of HfO_2, it seems that the oxidation is even more efficient probably due to the hygroscopic nature of the high-k layers [14] and the presence in the layer itself, after deposition, of OH groups in the layer.
Different strategies have been tested in order to avoid or limit this anomalous oxidation. The first one consists in co-implanting N in the HfO_2 layer in order to trap the oxygen atoms. For this, two N profiles have been implanted at very low energy, 1 and 2.5 keV with doses corresponding to 15 at.%. Higher N content would deteriorate the electrical properties [15]. The Si has been implanted at 3 keV with Si excess corresponding to 25 at.% as previously. We observe after annealing in the same conditions the formation of the same structure as previously, with amorphous white regions in the HfO_2 layer. The comparison of zero-loss EFTEM images with the same mean thickness as measured on the associated thickness map shows that these Si rich amorphous regions are present in lower proportion in the layer containing N (not shown). EFTEM image obtained by filtering at 23 eV (SiO_2 plasmon, see Fig. 4d) confirm that this region as well as the IL are SiO_2 made. These SiO_2 regions in the HfO_2 layer are again already present in the as-implanted sample. Furthermore, STEM-EELS experiments have shown that N is located near the Si substrate, in the SiO_2 IL (not shown). During annealing N has diffused from the film into the SiO_2 IL.
The IL is now 5.5 nm instead of 7.2 nm for the sample implanted with Si only, evidencing a lower oxidation of the Si substrate. A layer of NPs is now clearly visible in the BF images within the SiO_2 IL layer (see Fig. 4a). These particles are crystalline as we can see in the HREM image of Fig. 4b showing interplanar distances of 0.31 nm, which corresponds either

to Si or monoclinic HfO₂ (111) planes. EFTEM images at 17 eV (see Fig. 4c) confirm that these particles are composed of Si. In conclusion, N co-implantation has limited the anomalous oxidation of the implanted Si. Nevertheless, the Si NPs are still not formed in the high-k layer but again within the SiO_2 IL.

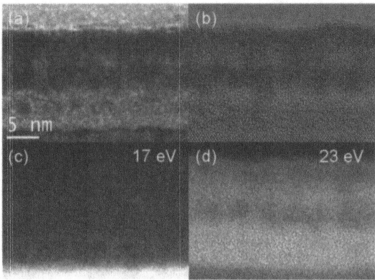

Fig. 4: (a) Bright field, (b) HREM, (c) EFTEM around the Si plasmon (17 eV) and (d) EFTEM around the SiO_2 plasmon (23 eV) for the N+Si implanted HfO₂ film.

The same strategy (N+Si implantation) has been used in thicker (9 nm) monoclinic HfO₂ layers as well has in 6 nm amorphous HfSiO layers, with the same result (oxidation of the Si remaining in the HfO₂ layer and formation after annealing of Si NCs within the IL layer). The annealing under NH₃ ambient instead of N₂ of Si implanted HfO₂ layers does not help in avoiding the Si oxidation, neither than the implantation through a SiN capping layer.

C-V measurements have been performed for the different implanted layers. The best memory window has been found when the anomalous oxidation of the Si implanted within HfO₂ leads to disconnected HfO₂ grains, aligned following 2 parallel layers (see Fig. 1b). In this case we have formed an unexpected structure made of HfO₂ quantum wells embedded within a SiO_2 insulating layer. Such a nanostructured layer (HfO₂ NCs within SiO_2) has already been fabricated by annealing HfSiO alloys under O₂ at high temperature and show nice electrical performances [16].

A third layer of discrete traps (Si NCs) is embedded in the SiO_2 IL separating the HfO₂ NCs from the Si substrate, giving rise to a third level NCs storage. In this case, significant hysteretic effect with a memory window (ΔV_{fb}) of 1.18 V is observed for sweeping voltages of ±6 V. It is noteworthy that effect of mobile ions on charge trapping can be ruled out, as no significant hysteresis has been observed in the control sample without any nanocrystal. The frequency dependent C-V curves were further measured in the range of 1 MHz to 50 kHz (not shown). Little frequency dispersion has been observed in the depletion region confirming that the charge storage was rather in NCs rather than in defect sites. All the samples show clockwise hysteresis indicating the movement of a net negative charge (electrons) in the dielectric layer. The most profound reason can be attributed to the electron injection from the gate, when the device is biased at accumulation, and subsequent trapping in the dielectric coating NCs. Further charge retention measurements are under process.

This unexpected but interesting structure will be optimised by the deposition of a control oxide in order to limit the charge leakages to the gate. Annealing under O_2 ambient could be used in order to better control the oxidation process as in ref. [16] and therefore the size and density of the HfO_2 NCs. Finally, HfO_2 layers are interesting for NCs memories mainly for their improved electrical characteristics when used as tunnel oxide. As a consequence, multilayers structures, composed of HfO_2 layers on Si substrate, with on top SiN layers, will be implanted in the future, to get rise from the material science difficulties in stabilising a Si phase in such dielectric layers.

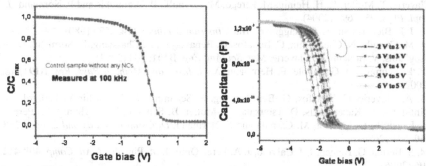

Fig. 5: High frequency capacitance-voltage (C-V) measurements of metal-insulator-semiconductor (MIS) structures for (a) the as-deposited HfO_2 layer and (b) the HfO_2 layer implanted with Si+ at 3 keV with an excess of 26 at.% and annealed at 1000°C for 30s.

CONCLUSION

The implantation of Si+ within HfO_2 based layers deposited by MOCVD on top of a Si substrate leads to the formation of Si NCs within the SiO_2 interfacial layer and to the oxidation of the Si implanted within the HfO_2 top layer. This anomalous oxidation of implanted Si in thin insulating layers is a well known process and is due to moisture penetration in the heavily damaged layers after ion implantation. Annealing under NH_3 ambient instead of N_2 has no detectable effect on the final structure and do not limit the oxidation extend. Nevertheless, this oxidation process is slightly reduced when co-implanting N before the Si implantation or when implanting through SiN capping layers. Finally, when the anomalous oxidation is the most efficient and leads to the clear separation of the HfO_2 grains, interesting memory characteristics are found.

ACKNOWLEDGEMENTS

This work is financed by the ANR project *"ANR/PNANO07-0053 – NOMAD"*.

REFERENCES

1. S. Tiwari, F. Rana, H. Hanafi, A. Hartstein, E. F. Crabbe and K. Chan, *Appl. Phys. Lett.* **68**, 1377 (1996)
2. H. I. Hanafi, S. Tiwari and I. Khan, *IEEE Trans. Electron Devices* **43**, 1553 (1996)

3. C. Bonafos, H. Coffin, S. Schamm, N. Cherkashin, G. Ben Assayag, P. Dimitrakis, P. Normand, M. Carrada, V. Paillard, A. Claverie, *Solid-State Electronics* **49**, 1734 (2005)

4. J. J. Lee, W. Bai and D.-L. Kwong, *IEEE 43rd Annual International Reliability Physics Symposium, San Jose* (2005)

5. J. J. Lee, X. Wang, W. Bai, N. Lu and D.-L. Kwong, *IEEE Trans. Electron Devices* **50**, 2067 (2003)

6. J. H. Chen, Y. Q. Wang, W. J. Yoo, Y.-C. Yeo, G. Samudra, D. SH Chan, A. Y. Du and D.-L. Kwong, *IEEE Trans. Electron Devices* **51**, 1840 (2004)

7. J. Lu, Y. Kuo, J. Yan and C.-H. Lin, *Jap. J. Appl. Phys.* **45**, L901 (2006)

8. M. Fanciulli, M. Perego, C. Bonafos, A. Mouti, S. Schamm, G.Benassayag, *Adv. Sci. Technol.* **51**, 156 (2006)

9. C. Bonafos, M. Carrada, N. Cherkashin, H. Coffin, D. Chassaing, G. Ben Assayag, A. Claverie, T. Müller, K. H. Heinig, M. Perego, M. Fanciulli, P. Dimitrakis and P. Normand, *J. Appl. Phys.* **95**, 5696 (2004)

10. J. P. Biersack and L. G. Haggmark, *Nucl. Instrum. Methods* **174**, 257 (1980)

11. M. Carrada, N. Cherkashin, C. Bonafos, G. Benassayag, D. Chassaing, P. Normand, D. Tsoukalas, V. Soncini, A. Claverie, Mat. *Sci. And Eng.* **B101**, 204 (2003)

12. B. Schmidt, D. Grambole, F. Herrmann, *Nucl. Instr. and Meth. in Phys. Res.* **B191**, 482 (2002)

13. A. Claverie, C. Bonafos, G. Ben Assayag, S. Schamm, N. Cherkashin,V. Paillard, , P. Dimitrakis, E. Kapetenakis, D. Tsoukalas, T. Muller, B. Schmidt, K. H. Heinig, M. Perego, M. Fanciulli, D. Mathiot, M. Carrada and P. Normand, *Diffusion in Solids and Liquids* **258-260**, 531 (2006)

14. S. Bernal, G. Blanco, J. J. Calvino, J. A. Pérez Omil, J. M. Pintado, *J. All. Comp.* **408-412**, 496 (2006)

15. M. Suzuki, A. Takashima, M. Koyama, R. Iijima, T. Ino, M. Takenaka, *Nucl. Instr. and Meth. in Phys. Res.* **B219-220**, 851 (2004)

16. Y.-H. Lin, C.-H. Chien, C.-T. Lin, C.-Y. Chang and T.-F. Lei, *IEEE Trans. Electron Devices* **53**, 782 (2006)

Mater. Res. Soc. Symp. Proc. Vol. 1160 © 2009 Materials Research Society   1160-H01-05

Performance Enhancement of TiSi$_2$ Coated Si Nanocrystal Memory Device

Huimei Zhou[1], Reuben Gann[2], Bei Li[1], Jianlin Liu[1] and J. A. Yarmoff[2]
[1]Department of Electrical Engineering, University of California, Riverside, California 92521
[2]Department of Physics and Astronomy, University of California, Riverside, California 92521

ABSTRACT

Self-aligned TiSi$_2$ coated Si nanocrystal nonvolatile memory was fabricated. This kind of MOSFET memory device is not only thermally stable, but also shows better performance in charge storage capacity, writing, erasing speed and retention characteristics. This indicates that CMOS compatible silicidation process to fabricate TiSi$_2$ coated Si nanocrystal memory is promising in memory device applications.

INTRODUCTION

The dimensions of Si based memory devices have approached the nanometer scale and Si nanocrystal embedded memory devices have significantly improved the performance of floating gate memory [1]. It was reported that the defects in Si nanocrystals boost up the long-term retention performance [2]. However, the defects based performance improvement is not stable in MOSFET memory device fabrication, in particular under subsequent high temperature annealing step [2]. New types of nanocrystal floating dots, such as double Si dots [3], Ge nanocrystals [4], metal [5-8] or metal-like [9] dots and dielectric nanocrystals (Al$_2$O$_3$, HfO$_2$, Si$_4$N$_3$, etc) [10-12], have been proposed to achieve memory devices with longer and stable retention performance. The higher defect levels in the dielectric require higher writing voltage, therefore inducing the erasing saturation [13]. A feasible solution to rule out the defect effect is to use nanocrystals with high density of states, such as metal nanocrystals [5-8], but the drawback of using Ge and metal nanocrystals is the inter-diffusion between nanocrystals and tunnel oxide during device integration [14-16]. The inter-diffusion degrades the tunnel oxide and worsens retention characteristics. Since the post annealing is necessary for most of the device process, the thermal stability of a memory cell has become very critical. In this work we proposed and experimentally verified a method to improve the thermal stability of the memory cell by using self aligned TiSi$_2$ coated Si nanocrystals technique. TiSi$_2$ coated Si nanocrystal memory can not only have faster writing/erasing speed, but also have longer retention performance than pure Si nanocrystal memory as a result of Fermi-level pinning effect and high density of states around the silicide.

EXPERIMENT

In device fabrication, first, 5 nm thermal oxide was grown in dry oxygen immediately followed by Si nanocrystal deposition. An ultra-thin (<0.5 nm) blanket Ti layer was then deposited. A modified two-step annealing for silicidation [17] was performed in nitrogen to coat the Si nanocrystals. The first annealing only forms silicide on Si nanocrystal and Ti on oxide remains as metal. After selectively removing the unreacted metal Ti, the wafer was annealed (second annealing) at 900 °C for 30 seconds to form thermally stable Si-rich silicide [18]. Control oxide of about 15 nm was then deposited, followed by a 350-nm-thick poly-silicon deposition in Low Pressure Chemical Vapor Deposition (LPCVD). After the formation of gate pattern, the poly-silicon gate and source/drain regions were implanted with phosphorus followed by dopant activation.

RESULTS AND DISCUSSION

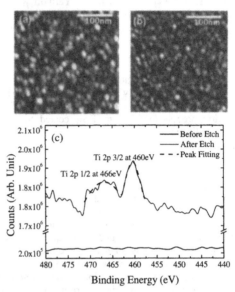

Fig. 1. AFM images of the TiSi$_2$ coated Si nanocrystals a) before and b) after diluted HF etching c) XPS result of the TiSi$_2$ coated Si nanocrystal sample before and after diluted HF etching

Figure 1(a) and (b) show the atomic force microscope (AFM) images of TiSi$_2$ coated Si nanocrystals before and after HF etching, respectively. The nanocrystal diameter is ~ 10 nm and the density is ~ 5×10^{11} cm^{-2}. The Si nanocrystals still exist after diluted HF etching, which indicates the fact that thin layer of Ti silicide has formed on the surface of the Si nanocrystals. Fig. 1(c) shows the results of X-ray photoelectron spectroscopy (XPS) of the same samples before and after HF etching. For the as-fabricated nanocrystal sample, there is one evident peak at 460 eV corresponding to Ti 2P3/2 states of TiSi$_2$. This Ti-related signal disappears after HF etching, which means the TiSi$_2$ portions of the nanocrystals were removed. The combination of AFM and XPS results suggests that TiSi$_2$ coated Si nanocrystals have been achieved.

To verify the details of coated shape of the nanocrystals, simple calculation was performed. A half circle shape of Si nanocrystal with height of 8nm and diameter of 16nm was proposed to deposit on SiO$_2$, as shown in Fig. 2 (a). Silicide process is suppressed by stress between SiO$_2$ and Si [19], here we assume the stress σ decreases as it is away from SiO$_2$:

$$\sigma_{nn} = A e^{\frac{B}{n^2}}.$$

Combining the concentration differential equation and stress related equation [20], the thickness of TiSi$_2$ is given by

$$\delta(THK) = \frac{\Delta t \times C \times k_s}{N_1}, \qquad k_s = k_{s0} \exp\left(-\frac{E_{k0} + \sigma_{nn} * V_{kp}}{k * T}\right)$$

Where Δt is the annealing time, C is the concentration of elements, K_S is the reaction rate, N_1 is the number of silicon atoms penetrating to the silicide layer, K_{S0}, E_{K0}, V_{KP} are parameters of the model. Fig 2 (b) shows the 2 dimension (2D) picture of TiSi$_2$ distribution in this structure. Si shows higher diffusion rate when it is away from the interface between Si and SiO$_2$. While at the edge of Si nanocrystal, slower diffusion rate results in the crescent shape of the coated TiSi$_2$.

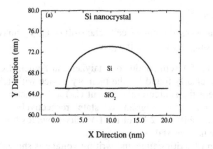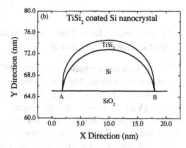

Fig. 2. a) Nanocrystal structure in the beginning of calculation, b) TiSi$_2$ signal distribution in 2D.

Figure 3(a) shows the schematic of TiSi$_2$ coated Si nanocrystal memory device. The blue color shows the Si nanocrystal and the orange ring covering the surface of Si nanocrystals represents TiSi$_2$ layer. Fig. 3(b) shows the 3D conduction band of TiSi$_2$ coated Si nanocrystal memory device. TiSi$_2$ layer which has lower energy level attaches the tunnel oxide and exists between control oxide and Si nanocrystals. Under writing, the electrons from Si substrate go through the tunnel oxide, Si nanocrystal and finally are confined in the TiSi$_2$ region.

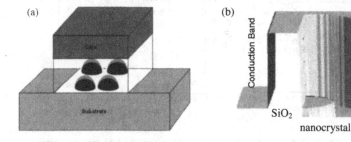

Fig. 3. (a) Schematic of TiSi$_2$ coated Si nanocrystal memory device, (b) Schematic of 3D conduction band structure of TiSi$_2$ coated Si nanocrystal memory device.

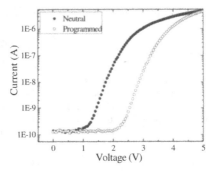

Fig. 4. Memory effect from TiSi$_2$ coated Si nanocrystal memory cell. The shift of I$_{ds}$-V$_g$ curve after writing operation indicates the electron storage in the floating gate.

The devices were characterized by HP 4155A semiconductor analyzer and Agilent LCR meter at room temperature. Memory effect was clearly found for the memory device with TiSi$_2$ coated Si nanocrystals, as shown in Fig.4, where the source-drain current (I$_{ds}$) as a function of gate voltage (Vg) is shown for the neutral state and programmed state, respectively. The programming was performed at 20 V for 1 s. The shift (~1.2 V) of the I-V curve towards higher gate voltage indicates the electron storage in the nanocrystals.

The threshold voltage shift as a function of writing time and writing voltage is shown in Fig. 5 (a) and Fig. 5 (b), respectively. It is found that the memory window becomes saturated as the writing time elapses with Vg fixed at 15 V. This is due to the Coulomb blockade effect caused by the small nanocrystal size. The saturated memory window for the device with TiSi$_2$ coated Si nanocrystals is higher than the Si nanocrystal memory device. The saturation level increases with writing voltage which is shown in Fig. 5 (b). It is interesting to note in Fig.5 (b) that the threshold voltage shift exhibits an obvious difference between TiSi$_2$ coated Si

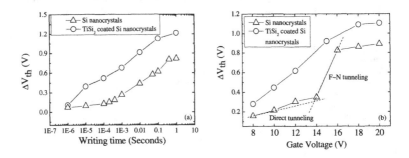

Fig. 5. Threshold voltage shift as a function of a) writing time and b) writing voltage, for memory cells with TiSi$_2$ coated Si nanocrystals and reference Si nanocrystals.

14

nanocrystal and Si nanocrystal memory devices. This is attributed to the different charge injection mechanisms in the two kinds of devices. In $TiSi_2$ coated Si nanocrystal case, $TiSi_2$ layer attaches the tunnel oxide and has lower energy level compared to the reference Si nanocrystals. The charge injection has already established through Fowler–Nordheim (F-N) tunneling at the voltage around 8V. $TiSi_2$ coated Si nanocrystals have more charges to be stored in the silicide because of its higher density of state (DOS). In Si nanocrystals, the energy levels are higher than that of $TiSi_2$ layer and the tunneling is direct tunneling at the voltage below 14V and F-N type at the voltage larger than 14V. As writing voltage increases further (>16 V), both devices become saturated.

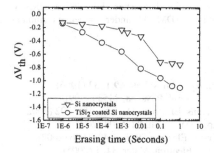
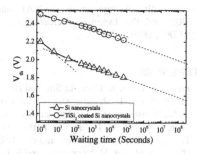

Fig. 6. Threshold voltage shift as a function of erasing time for memory cells with $TiSi_2$ coated Si nanocrystals and reference Si nanocrystals.

Fig. 7. Retention performance comparison between a reference Si nanocrystal and $TiSi_2$ coated Si nanocrystal memory devices. The writing was done at 20 V for 1s.

The threshold voltage shift as a function of erasing time at the erasing voltage of -15V is shown in Fig.6. The erasing is found speeding up with the increase of the erasing time in both samples. $TiSi_2$ coated Si nanocrystal memory shows higher erasing speed. In the $TiSi_2$ coated Si nanocrystal memory device, two factors make the erasing speed faster than the Si nanocrystal case: First, crescent shape of $TiSi_2$ layer makes the point discharge possible in the sharp area of the $TiSi_2$ layer. Second, tunnel oxide endures larger potential drop in the $TiSi_2$ coated Si nanocrystal, which helps electrons to go through by F-N tunneling.

The retention characteristics are shown in Fig. 7 for the two devices with $TiSi_2$ coated Si nanocrystals and reference Si nanocrystals, respectively. These devices were programmed at 20 V, with an initial memory window of ~1.27V and 0.9V respectively. $TiSi_2$ coated Si nanocrystal memory shows slower charge loss rate in earlier retention stage. In this case, the electrons stay in the $TiSi_2$ channel and most of them have the lower energy level which is difficult to go through the tunnel oxide layer. The other reason is that the interface between the $TiSi_2$ and tunnel oxide has very small area which is blocked by the coulomb blockade effect and results in the difficulty of leakage. After the extrapolation of the curves to 10 years at room temperature, the remaining memory window is predicted to be 0.65 V, which means 52% charge left, while that is only 0.25 V, which means only 28% charges left for the reference Si nanocrystal memory device, as can be seen from Fig. 7.

CONCLUSIONS

In summary, we have successfully fabricated TiSi$_2$ coated Si nanocrystal devices by self-aligned silicidation method. Due to TiSi$_2$ induced lower energy levels, the charge storage occurs mainly in the TiSi$_2$ layer. Therefore, as compared to the reference Si nanocrystal memory, the memory device with TiSi$_2$ coated Si nanocrystals shows a larger charge storage capacity, higher writing and erasing speed and much better retention performance.

ACKNOWLEDGMENTS

The authors acknowledge the financial and program support of FCRP center on Function Engineered NanoArchitectonics (FENA), and the National Science Foundation (ECCS-0725630), and the Defense Microelectronics Activity (DMEA) under agreement number H94003-09-2-0901.

REFERENCES

1. S. Tiwari, F. Rana, K. Chan, L. Shi, and H. Hanafi, Appl. Phys. Lett. **69**,1232 (1996).
2. Y. Shi, K. Saito, H. Ishikuro, and T. Hiramoto, Jpn. J. Appl. Phys. **38**, 2453 (1999).
3. R. Ohba, N. Sugiyama, K. Uchida and etc., IEEE Trans. Electron Devices **49**, 1392 (2002).
4. Q. Wan, C. L. Lin, W. L. Liu, and T. H. Wang, Appl. Phys. Lett. **82**, 4708 (2003).
5. Z. T. Liu, C. Lee, V. Narayanan, and etc., IEEE Trans. Electron Devices **49**, 1606 (2002).
6. C. H. Lee, J. Meteer, V. Narayanan, and E. C. Kan, J. Electron. Mater. **34**, 1 (2005).
7. J. J. Lee and D. L. Kwong, IEEE Trans. Electron Devices **52**, 507 (2005).
8. T. C. Chang, P. T. Liu, S. T. Yan and S. M. Sze, Electrochem. Solid-State Lett. **8**, G71 (2005).
9. S. Choi, S. S. Kim, M. Chang, H. S. Hwang, and etc., Appl. Phys. Lett. **86**, 123110 (2005).
10. J. H. Chen, W. J. Yoo, D. S. H. Chan, and L. J. Tang, Appl. Phys. Lett. **86**, 073114 (2005).
11. Y. H. Lin, C. H. Chien, C. T. Lin, and etc., IEEE Electron Device Lett. **26**, 154 (2005).
12. S. Y. Huang, K. Arai, K. Usami, and S. Oda, Nanotechnology **3**, 210 (2004).
13. I. De. Wolf, D. J. Howard, A. Lauwers, K. Maex, and etc., Appl. Phys. Lett. **70**, 2262 (1997).
14. T. H. Ng, W. K. Chim, W. K. Choi, V. Ho and etc., Appl. Phys. Lett., **84**, 4385 (2004).
15. Ya-Chin King, Tsu-Jae King, and Chenming Hu, IEDM Tech. Dig. Page 115 (1998).
16. Tae-Sik Yoon, Jang-Yeon Kwon, Dong-Hoon Lee and etc., J. Appl. Phys. **87**, 2449(2000).
17. Y. Zhu, D. T. Zhao, R. G. Li, and J. L. Liu, Appl. Phys. Lett. **88**, 103507 (2006).
18. J. P. Gambino, and E. G. Colgan, Material Chemistry and Physics **52**, 99 (1998).
19. Victor Moroz and Takako Okada, Mat. Res. Soc. Symp. Vol. 611 (2000).
20. P. Fornara, S Denorme, E. de Berranger and etc., Microelectronics Journal, 29, 71-81 (1998).
21. Y. Zhu, D. Zhao, R Li, and J. Liu, Journal of Applied Physics 97, 034309 (2005).

Mater. Res. Soc. Symp. Proc. Vol. 1160 © 2009 Materials Research Society   1160-H01-09

VSS-Induced NiSi$_2$ Nanocrystal Memory

Bei Li and Jianlin Liu
Quantum Structures Laboratory, Department of Electrical Engineering, University of California, Riverside, California 92521

ABSTRACT

NiSi$_2$ nanocrystals were synthesized and used as the floating gate for nonvolatile memory application. Vapor-solid-solid mechanism was employed to grow the NiSi$_2$ nanocrystals by introducing SiH$_4$ onto the Ni catalysts-covered SiO$_2$/Si substrate at 600°C. The average size and density of the NiSi$_2$ nanocrystals are 7~10nm and 3×10^{11} cm^{-2}, respectively. Metal-oxide-semiconductor field-effect-transistor memory with NiSi$_2$ nanocrystals was fabricated and characterized. Programming/erasing, retention and endurance measurements were carried out and good performances were demonstrated.

INTRODUCTION

Silicide nanocrystals are believed as one of the promising candidates to replace traditional Si as the floating gates in nonvolatile memory. The time-voltage dilemma prevents conventional Si memory scale beyond 32nm technology node. Using silicide nanocrystals may extend this scale limit by enhancing the device retention without compromising the program efficiency. Because silicides are metallic materials, the strong coupling between the channel and floating gates helps improve the programming speed and deeper quantum well formed between the metals (those Fermi-levels within the energy band gap of Si), and Si substrate elongates the retention time. The storage capability can also be improved using silicide nanocrystals due to their high density of states. The large memory window makes it possible for the device to be used for multi-bit applications. Another advantage for silicide materials is their thermal stability. The current long retention reported in Si nanocrystal memory is mainly due to the charging on the defect levels within Si, which is not thermally robust.

Various silicide nanocrystals [1-5] have been reported showing good memory performance. The methods include ebeam evaporation, sputtering with post rapid thermal annealing (RTA). In this work, we use a novel method, i.e., Vapor-Solid-Solid (VSS), which is mainly used for nanowires growth [6-8], to synthesize silicide nanocrystals.

EXPERIMENT

A thin layer of metal Ni was deposited by ebeam evaporation on a 5nm thermally grown SiO$_2$ covered p-Si (100) substrate. In-situ annealing in N$_2$ at 900°C for 30s was carried out to release the stress and densify the oxide film. The wafer was then transferred to a low vapor chemical vapor deposition (LPCVD) furnace for silicide synthesis. As the furnace temperature increases to the growth temperature, ~600°C, Ni nanocrystals were formed, then Si precursor (SiH$_4$) was introduced to diffuse into/react with Ni to form silicide. The growth time was calibrated and the SiH$_4$ gas was shut off at the time when no Si nanowires growth underneath the catalysts. Control oxide of 25nm was deposited on the nanocrystals followed by 350nm

17

polysilicon gate formation. Phosphorous was implanted to make heavily doped source/drain/gate regions. Aluminum was evaporated and patterned as the contacts to complete the fabrication. The device feature is 1μm.

RESULTS and DISCUSSION

Figure 1 shows the atomic force microscopy (AFM) image of $NiSi_2$ nanocrystals. The nanocrystal size is determined to be 7~10nm and the density is $3\times10^{11}cm^{-2}$. X-ray photoelectron spectroscopy (XPS) was used to ascertain the chemical composition of the nanocrystals to be $NiSi_2$ (Results are not shown here).

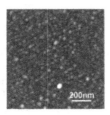

Figure 1 AFM image of $NiSi_2$ nanocrystals on SiO_2/Si substrate.

Figure 2 (a) shows the typical capacitance-voltage (C-V) sweep characteristics of $NiSi_2$ nanocrystals metal-oxide-semiconductor (MOS) memory. Nanocrystals are embedded between 5nm tunnel oxide and 25nm control oxide. The contact materials are aluminum. Three different scanning range, ± 5V, ± 8V, ± 10 V, have been applied to monitor the flat band voltage shifts as gate voltage sweeps. It is found that the memory window increases as scan voltage increases. With ± 5V scanning, memory window is 0.268V and with ± 10V scanning, memory window is 1.052V. Figure 2 (b) shows the sweep of a reference sample, where no nanocrystals are between control and tunnel oxide. Insignificant flat band voltage shift is observed, indicating the memory effect in $NiSi_2$ nanocrystal MOS memory is due to the electron charging in the nanocrystals, rather than interface/defect levels in the dielectric layers.

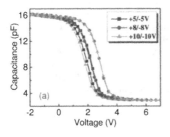
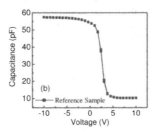

Figure 2 Capacitance-voltage (C-V) sweep of (a) MOS memory with $NiSi_2$ nanocrystals and (b) reference MOS device without any nanocrystals.

NiSi$_2$ nanocrystal metal-oxide-semiconductor field-effect-transistor (MOSFET) memory was fabricated using standard n-MOSFET process and the characterizations were carried out using Agilent 4155C semiconductor parameter analyzer to sense the threshold voltage and Agilent 81104A pulse/pattern generator to program/erase the memory. Figure 3 shows the charge retention performance of NiSi$_2$ nanocrystal memory after Fowler-Nordheim (FN) programming and erasing. The program and erase conditions are gate voltages of +20V and -20V, respectively, for 5 seconds. Retention performance at both programmed and erased states were recorded. After 10^5 seconds, the threshold voltage shifts change from 2.72V to 1.48V at programmed state and from -1.52V to -1.18V at the erased state, respectively. Ten-year retention is indicated in the figure to show that our device can survive ten years with memory window open. Figure 4 shows the endurance characteristics of the NiSi$_2$ nanocrystal memory. The device was charged and discharged by FN with V$_G$=18V/20ms and V$_G$=-18V/200ms, respectively. Up to 10^5 times of operation, the memory window shrinks from 1.749V to 1.351V, i.e. the charge loss after 10^5 time of programming/erasing is only ~23%.

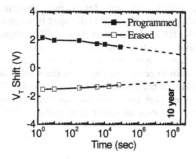

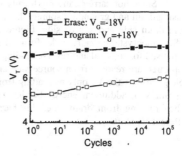

Figure 3 Retention characteristics of NiSi$_2$ NC memory at programmed and erased states

Figure 4 Endurance characteristics of NiSi$_2$ NC memory

In addition to FN, we also tried hot carrier injection to program the device, where both gate and drain were biased. As electrons transfer from source to drain and once the energy they gain can overcome the tunnel oxide barrier, they can charge and store in the floating gate, only around drain side. Shown in Fig. 5 are the transient programming characteristics of NiSi$_2$ nanocrystal memory. Figures 5 (a) and (b) show the drain bias and control gate bias effect on the programming efficiency, respectively. As increasing of either drain voltage or gate voltage, the programming gets faster and more charges are stored in the floating gate.

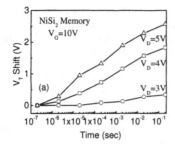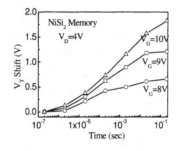

Figure 5 Hot carrier injection programming characteristics of NiSi$_2$ nanocrystal memory. (a) drain bias and (b) control gate bias dependence of the programming efficiency.

Since hot carrier can only locally inject to the floating gate, for example, to drain side if biasing drain and grounding source, the threshold voltage (V_T) shift is different between reading from drain side and source side. Figure 6 is the retention characteristics of HCI-programmed NiSi$_2$ nanocrystal memory under forward and reverse reading conditions. Three reading voltages were used for forward and reverse reading, therefore six retention curves are shown in the figure. Top three are reading from source side and bottom three are reading from drain side. Reading voltage affects the V_T only as reading from drain side because after HCI programming (biasing V_G and V_D), additional energy barrier was formed near the drain side for channel electrons. When reading from drain side, this barrier was lowered at larger read voltages, leading to different V_T shift.

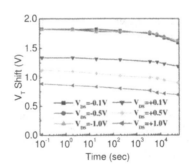

Figure 6 Retention characteristics of HCI-programmed NiSi$_2$ nanocrystal memory. Forward and reverse reads were used to sense the V_T shift.

20

CONCLUSIONS

Ni-catalyzed NiSi$_2$ nanocrystals were synthesized on thin oxide layer via VSS mechanism. Optimized conditions were obtained to get the silicide nanocrystals. MOSFET memory with NiSi$_2$ nanocrystals as the floating gate were characterized and good performance in terms of programming/erasing speed, long retention time and acceptable endurance were demonstrated.

ACKNOWLEDGMENTS

The authors acknowledge the financial and program support of the Microelectronic Advanced Research Corporation (MARCO) and its Focused Center on Function Engineered NanoArchitectonics (FENA), the National Science Foundation (ECCS-0725630) and the Defense Microelectronics Activity (DMEA) under agreement number H94003-09-2-0901.

REFERENCES

1. C. W. Hu, T. C. Chang, P. T. Liu, C. H. Tu, S. K. Lee, S. M. Sze, C. Y. Chang, B. S. Chiou, and T. Y. Tseng, Appl. Phys. Lett. **92**, 152115 (2008)
2. J. Kim, J. Yang, J. Lee, and J. Hong, Appl. Phys. Lett. **92**, 013512 (2008)
3. F. M. Yang, T. C. Chang, P. T. Liu, Y. H. Yeh, Y. C. Yu, J. Y. Lin, S. M. Sze, and J. C. Lou, Thin Solid Films **516**, 360 (2007)
4. W. R. Chen, T. C. Chang, J. L. Yeh, S. M. Sze, and C. Y. Chang, J. Appl. Phys. Lett. **104**, 094303 (2008)
5. Y. Zhu, B. Li, J. L. Liu, G. F. Liu, and J. A. Yarmoff, Appl. Phys. Lett. **89**, 233113 (2006)
6. Y. W. Wang, V. Schmidt, S. Senz, and U. Goesele, Nature Nano. **1**, 186 (2006)
7. J. L. Lensch-Falk, E. R. Hemesath, D. E. Perea, and L. J. Lauhon, J. Mater. Chem. **19**, 849 (2009)
8. H. Nordmark, H. Nagayoshi, N. Matsumoto, S. Nishimura, K. Terashima, C. D. Marioara, J. C. Walmsley, R. Holmestad, and A. Ulyashin, J. Appl. Phys. Lett. **105**, 043507 (2009)
9. J. Stumper, H. J. Lewerenz, and C. Pettenkofer, Phys. Rev. B **41**, 1592 (1990).
10. F. J. Grunthaner, P. J. Grunthaner, R. P. Vasquez, B. F. Lewis, and J. Maserjian, J. Vac. Sci. Technol. **16**, 1443 (1979).
11. J. F. Moulder, W. F. Stickle, P. E. Sobol, and K. D. Bomben, Handbook of X-Ray Photoelectron Spectroscopy. Physical Electronics: Eden Prairie MN (1995).
12. T. T. A. Nguyen and R. Cinti, Physica Scripta. T4, 176 (1983).
13. Schmid, P. E.; Ho, P. S.; Tan, T. Y. J. Vac. Sci. Technol. **1982**, 20 (3), 688-689.
14. Nash, P. Phase Diagrams of Binary Nickel Alloy, Monograph Series on Alloy Ahase Diagram, 6, ASM International, The Material Information Society USA **1992**.

Charge Trap Memory I

Mater. Res. Soc. Symp. Proc. Vol. 1160 © 2009 Materials Research Society 1160-H02-01

Reliability of nc-ZnO Embedded ZrHfO High-*k* Nonvolatile Memory Devices Stressed at High Temperatures

Chia-Han Yang[1, 2], Yue Kuo[1], Chen-Han Lin[1] and Way Kuo[3]

[1]Thin Film Nano & Microelectronics Research Laboratory, Texas A&M University,
College Station, TX 77843-3122, U.S.A.
[2]Department of Industrial and Information Engineering, University of Tennessee,
Knoxville, TN 37996, U.S.A.
[3]City University of Hong Kong, Hong Kong

ABSTRACT

The nanocrystalline ZnO embedded Zr-doped HfO$_2$ high-*k* dielectric has been made into MOS capacitors for nonvolatile memory studies. The device shows a large charge storage density, a large memory window, and a long charge retention time. In this paper, authors investigated the temperature effect on the reliability of this kind of device in the range of 25°C to 175°C. In addition to the trap-assisted conduction, the memory window and the breakdown strength decreased with the increase of the temperature. The high-k film's conductivity increased and the nc-ZnO's charge retention capability decreased with the increase of temperature. The nc-ZnO retained the trapped charges even after the high-k film broke down and eventually lost the charges at a higher voltage. The difference between these two voltages decreased with the increase of the temperature.

INTRODUCTION

The evolution of the MOSFET technology has been driven by the aggressive shrinkage of the device size and at the same time, to improve the performance and to increase the circuit density. However, when the thickness of the gate SiO$_2$ is reduced from 3.5nm to 1.5nm, the leakage current at a gate bias of 1V increases drastically from 10^{-12} A to 10 A due to quantum-mechanical tunneling [1]. Currently, there are many researches on replacing SiO$_2$ with a high dielectric constant (high-*k*) material, such as Si$_3$N$_4$, HfSi$_x$O$_y$, HfO$_2$, and ZrO$_2$, in order to achieve a low leakage current with improved reliability [2]. High-*k* dielectrics are also required for the nanosize nonvolatile memory (NVM) devices [1]. The conventional poly-Si floating-gate structure, which includes a continuous poly-Si thin film in the SiO$_2$ gate dielectric layer to retain charges, is prone to lose all charges when a single leakage path is formed. When the continuous poly-Si layer is replaced with discrete nanodots, the above problem can be eliminated because one leaky path in the tunnel oxide can only drain charges stored in a few nanodots [3]. When the SiO$_2$ tunnel dielectric layer is replaced with a high-*k* film, the leakage paths are difficult to form because the larger physical thickness. Due to the low band offset between the high-*k* film material and silicon (Si), this kind of memory device requires a low operating power. Therefore, the nanocrystals embedded high-*k* structure can potentially replace the conventional poly-Si floating-gate structure in high-density NVMs [3-5]. The Zr-doped HfO$_2$ (ZrHfO) high-k film has many superior material and electrical properties than the undoped HfO$_2$ film in areas such as the high crystallization temperature, the large effective *k* value, and the low interface state density

[6-10]. Recently, nanocrystalline (nc) metals, metal oxides or semiconductor materials, such as Ru, Ni, indium tin oxide (ITO), Si, and zinc oxide (ZnO), have been embedded into the high-k dielectric for nonvolatile memory applications. They provide more choices on work functions, nanocrystal sizes, deposition methods, etc. [11-14]. The nc-ZnO embedded high-k gate dielectric should have excellent electron retention characteristics based on the band alignment between nc-ZnO and Si [15]. It has been demonstrated that the ZrHfO/nc-ZnO /ZrHfO tri-layer structure can trap a large number of electrons with a long retention time [11]. Yang et al. [16] also reported that the nc-ZnO embedded ZrHfO film had a large charge holding capability with strong breakdown strength. However, most reliability studies on nonvolatile memories were done at the room temperature [16-17]. The influence of temperature on the memory function is unknown. In this paper, authors investigated the temperature influence on the retention characteristics and the breakdown mechanism of the nc-ZnO embedded ZrHfO MOS capacitor.

EXPERIMENT

The nc-ZnO embedded ZrHfO thin film was prepared into a MOS capacitor, as shown in Figure 1, for reliability studies. The bottom ZrHfO tunnel oxide/ZnO/top ZrHfO control oxide tri-layer was sputter deposited sequentially in the same chamber with a one pumpdown process without breaking the vacuum on the HF pre-cleaned p-type Si (100) wafer (doping concentration at 10^{15} cm^{-3}). The ZrHfO films were deposited using a Hf/Zr (88:12 wt%) composite target in an Ar/O$_2$ (1:1) mixture at 5 mTorr at 60W and room temperature for 2 min (for the tunnel oxide) and 4 min (for the control oxide), separately. The ZnO film was sputter deposited from the Zn target in Ar/O$_2$ (1:1) at 10 mTorr and 60W. The as-deposited amorphous ZnO layer was transformed into nanocrystals after post-deposition annealing (PDA) at 800°C for 180 sec under N$_2$ with rapid thermal annealing. An aluminum (Al) film was sputtered on the high-k stack surface and wet etched to form the gate electrode with an area of 7.85x10^{-5} cm^2. The backside of the wafer was deposited with Al for ohmic contact. The equivalent oxide thickness (EOT) of the embedded high-k stack was 7.8 nm estimated from the C-V curve at 1MHz using an HP4284A. The average crystal size of the nc-ZnO was about 3.0 nm determined from the full width at half-maximum of the ZnO (100) peak at 2θ=31° using the Scherrer equation. The discretely dispersed nc-ZnO was observed from TEM pictures [11]. The capacitor showed a counterclockwise C-V hysteresis with a flat band voltage shift of 1.22V under the gate sweep range of -6V$\rightarrow$+6V$\rightarrow$-6V [11]. The non-embedded ZrHfO sample showed negligible C-V hysteresis.

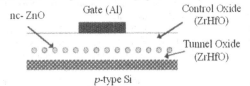

Figure 1. Cross-sectional view of the nc-ZnO embedded MOS capacitor.

RESULTS AND DISCUSSION

Figure 2 shows J–V curves of the nc-ZnO embedded capacitor under different temperatures in the log-lin scale. The gate was swept from 0 to -5V in Fig. 2(a) and 0 to +5V in

Fig. 2(b). Below -1.5V, a hole-rich accumulation is gradually formed; above 1.5V, an electron-rich inversion layer is gradually established. The tunneling current rapidly increases at a large gate voltage (V_g). Both figures show that the leakage current density increases slightly with the increase of temperature, which is consistent with the trap-assisted tunneling (TAT) mechanism, such as Frenkel-Poole tunneling or Schottky emission [24,25]. For the SiO_2 dielectric, the Fowler–Nordheim (F–N) tunneling occurs under the high electric field. However, for the HfO_2 high-k dielectric, the F-N tunneling mechanism could exist at a low electric field condition [26, 27]. This is because the barrier height between HfO_2 and Si is much lower than that between SiO_2 and Si, i.e., 1.5 eV vs. 3.5 eV [26,27]. Since the band gap of ZrHfO is similar to that of HfO_2, the F-N tunneling may also exist in the nc-ZnO embedded ZrHfO sample [28]. In addition, since the hole barrier height between ZrHfO and Si is larger than the electron barrier height, i.e., 3.4 eV vs. 1.5 eV, the leakage current density in Fig. 2(a) is smaller than that in Fig. 2(b) under the same magnitude of V_g.

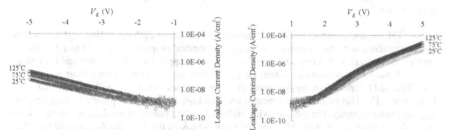

Figure 2. *J-V* curves of the nc-ZnO embedded MOS capacitor measured (a) from 0V to -5V and (b) from 0V to +5V.

The charge retention efficiency of the nc-ZnO embedded capacitor was studied with the following method [11, 12]. First, a V_g was applied to the capacitor for a period of time to trap charges to the dielectric layer, which is the "write" step. Second, after releasing the V_g, the *C-V* curve was measured in a small V_g range, i.e., -2V to +1V. Only negligible charges were injected into or removed from the capacitor during the *C-V* measurements because of the small V_g range. The flat band voltage, V_{FB}, calculated from the *C-V* curve reflects the capacitor's charge retention state. Third, the *C-V* measurement step was repeated every 1000s.

The flat band voltage shift (ΔV_{FB}), which is defined as V_{FB} (after stress release) - V_{FB} (original, before stress), can be expressed as a function of the retention time (t), as shown in Figure 3. The electron retention characteristics were measured at different temperatures after the +6V, 90s "write" step. The magnitude of the charge storage decreased with the increase of temperature. This can be attributed to the reduction of electron trapping and retention capacities of the nc-ZnO layer due to the increased electron thermal energy at elevated temperature [18]. In addition, the dielectric's conductance increases with the increase of temperature [30], which also can contribute to the leakage of the stored charges to the Si wafer. After 10,000 s, the capacitor lost 22.8%, 28.1% and 38% of originally stored charges at 25°C, 75°C and 125°C, respectively. The result shows that the nc-ZnO embedded ZrHfO high-k sample can retain charges for more than 10 years at 25°C, e.g., with ΔV_{FB} = 0.3V. However, the lifetime of the sample decreased to 2 years at 75°C and 1.5 months at 125°C, separately.

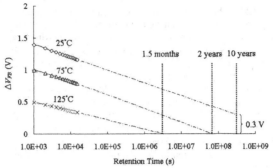

Figure 3. Charge retention characteristic of nc-ZnO embedded high-k thin film at different temperatures.

The thermal effect of breakdown mechanism for the nc-ZnO embedded high-k thin film can be examined with the ramp-relax measurement method proposed in refs. 19 and 31. To perform this experiment, a negative gate voltage ($-V_g$) is applied to the capacitor and the leakage current density J_{ramp} is measured. Then, the gate bias voltage is removed for a short period of time followed by immediate measurement of the relaxation current J_{relax} with a small gate voltage of 0.1V. This procedure is repeated until the high-k stack is completely broken, i.e., the J_{ramp} increases abruptly. For a metal oxide high-k film, the polarity of the J_{relax} is opposite to that of the J_{ramp} before it is broken. After the capacitor is broken, the J_{relax} and J_{ramp} should have the same polarity [20]. This phenomenon, i.e., the polarity change of the J_{relax}, has been successfully used to detect the breakdown sequence of the ultra thin high-k film [19,20,21,31]. Here, authors applied the same ramp-relax method to investigate the breakdown phenomena of nc-ZnO embedded high-k capacitors at different temperatures. Figure 4 shows (a) J_{ramp}-V_g and (b) J_{relax}-V_g curves of capacitors at 25°C, 75°C, 125°C and 175°C. The 25°C J_{ramp}-V_g curve is composed of three sections. First, the J_{ramp} increases slowly and smoothly with the increase of $-V_g$ from 0V to about -5V. In this region, charges start to trap to the embedded high-k stack and at the same time, the spot-connected breakdown path gradually forms [22]. In the second section, the quasi-breakdown initiates where spot-connected paths begin to be realized [22-23]. In this section, the leakage current increases faster with the increase of $-V_g$ due to the existence of a small number of the connected paths. In the third section, the complete breakdown occurs when the film becomes conductive and the J_{ramp} suddenly becomes very large [22]. Compared with the 25°C J_{ramp}-V_g curve, the smooth transition of the curve in section 1 at 75°C or 125°C is not obvious, i.e., ends before -2 volt. The short period of the smooth section shows that the increase of temperature provides enough energy for charges to form the spot-connected path fast.

The breakdown strength V_{BD} of the capacitor can be identified as the V_g corresponding to the abrupt jump of the J_{ramp}. Yang et al [16, 17] demonstrated that the nc-ZnO embedded ZrHfO sample had a larger breakdown voltage than the non-embedded ZrHfO sample. Since ZnO is semiconductor, it is more difficult to break than the surrounding dielectric material. The failure of the nc-ZnO embedded high-k film at room temperature is from the breakdown of the bulk ZrHfO [16, 17]. Fig. 4 (a) further shows that the breakdown strength of the nc-ZnO embedded ZrHfO sample decreases with the increase of the temperature. When $-V_g$ is small, it is very

28

difficult to form an electron-rich layer at the high-k/Si interface with enough energy to tunnel through the dielectric layer. Therefore, the J_{ramp}-V_g and J_{relax}-V_g curves are temperature independent at the small -V_g. When -V_g is large enough, an electron-rich layer is gradually formed with the increase of the voltage. When enough energy is obtained, they tunnel through the dielectric layer following the temperature-dependent F-P or Schottky emission mechanism.

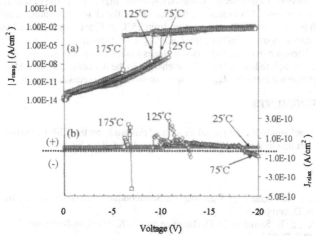

Figure 4. (a) J_{ramp}-V_g and (b) J_{relax}-V_g curves of nc-ZnO embedded ZrHfO capacitor at different temperatures.

The nc-ZnO effect on the J_{relax}-V_g curve is clearly shown in the new Fig. 4(b). At 25°C, the polarity of J_{relax} is opposite to that of J_{ramp} before and after the breakdown of the high-k film. This is because that the nc-ZnO does not lose its charge trapping or retention capability when the high-k part of the film breaks down. As the temperature increases, charges are easier to tunnel through the high-k film and trapped to the nc-ZnO site. Therefore, before the high-k film breaks down, the J_{relax} increases with the increase of temperature, as shown in Fig. 4(b). When the high-k film breaks, the whole nc-ZnO embedded film behaves like a conductor. The conductivity increases with the increase of temperature. In this case, the measured J_{relax} is the leakage current, which has an opposite polarity as that of the before breakdown. Therefore, the nc-ZnO loses the charge-retention capability at a V_g larger than the breakdown voltage of the bulk high-k film. The difference between these two V_g's decreases with the increase of the temperature. Fig. 4(b) shows that at 75°C, the J_{ramp}-V_g curve breaks at -10.1 V while the polarity change in the J_{relax}-V_g curve occurs at -18.3 V. At 125°C, the J_{ramp}-V_g curve breaks at -9.2 V while the polarity change in the J_{relax}-V_g curve occurs at -12.7 V. Eventually, at 175°C, the J_{ramp}-V_g curve breaks at the same V_g as that of the polarity change in the J_{relax}-V_g curve, i.e., -6.8 V. At 175°C, the nc-ZnO loses its charge-trapping capability at the bulk high-k breakdown voltage. In addition, the abruptness of the change of the polarity increases with the increase of temperature because the loss of charge retention and the breakdown of the high-k film almost occur at the same time.

CONCLUSIONS

Temperature effects on the charge retention characteristics and breakdown mechanism of the nc-ZnO embedded high-k dielectric have been investigated. The charge storage capacity decreased with the increase of temperature due to the increase of thermal energy of the trapped charge and the increase of the electric conductivity of the high-k film. The charge retention vs. time curve was measured. After releasing the gate stress, the charge retention capability decreased with the increase of the temperature from 25°C to 175°C. At the same temperature, the loss of charge retention of nc-ZnO occurred at a larger -V_g than the breakdown voltage. However, the difference between these two voltages reduced with the increase of the temperature. the complete loss of the trapped charges in the nanocrystals embedded high-k film can only be determined from the J_{relax}-V_g curve not the J_{ramp}-V_g breakdown curve.

ACKNOWLEDGMENTS

Authors acknowledge the partial support of this study by the NSF CMMI-0654172 project. C.-H. Lin thanks Applied Materials for providing fellowship.

REFERENCES

1. The International Technology Roadmap for Semiconductors. Semiconductor Industry Association, December 2003.
2. S. Chatterjee, S. K. Samanta, H. D. Banerjee and C. K. Maiti, in *Semicond. Sci. Technol.*, vol. 17, p. 993, 2003.
3. S. Tiwari, F. Rana, H. Hanafi, A. Hartstein, E. F. Crabbé, and K. Chan, in *Appl. Phys. Lett.*, vol. 68, no. 10, p. 1377, 1996.
4. B. De Salvo, G. Ghibaudo, G. Pananakakis, P. Masson, T. Baron, N. Buffet, A. Fernandes and B. Guillaumot, in *IEEE Trans. on Electron Devices*, vol. 48, p. 1789, 2001.
5. J. De Blauwe, in *IEEE Trans. Nanotechnol.*, vol. 1, p. 72. 2002.
6. J. Yan, Y. Kuo, and J. Lu, in *Electrochem. Solid-State Lett.*, vol. 10, H199, 2007.
7. Y. Kuo, J. Lu, S. Chatterjee, J. Yan , T. Yuan , H.- C. Kim, W. Luo, J. Peterson and M. Gardner, in *ECS Trans*, vol. 1 (5), p. 447, 2006.
8. D. H. Triyoso, in *ECS Trans.*, vol. 3 (3), p. 463, 2006.
9. Y. Kuo, in *ECS Trans.*, vol. 3 (3), p. 253, 2006.
10. Y. Kuo, in *ECS Trans.*, vol. 2 (1), p. 13, 2006.
11. J. Lu, C.-H. Lin and Y. Kuo, in *JES*, vol. 115(6), H386, 2008.
12. A. Birge and Y. Kuo, in *JES*, vol. 154 (10), H887, 2007.
13. D. B. Farmer and R. G. Gordon, in *J. Appl. Phys.*, vol. 101, 124503, 2006.
14. J. J. Lee, Y. Harada, J. W. Pyun and D. L. Kwong, in *Appl. Phys. Lett.*, vol. 86, 103505, 2005.
15. D. G. Baik and S. M. Cho,in *Thin Solid Film*, vol. 354, p. 227, 1999.
16. C. H. Yang, Y. Kuo, C. H. Lin, R. Wan and W. Kuo, in *Mat. Res. Soc. Symp. Proc.*, 1071, F02-09, 2008.
17. C. H. Yang, Y. Kuo, C. H. Lin, R. Wan and W. Kuo, in *Intl. Rel. Phys. Symp.*, p. 46, 2008.
18. J. H. Chen, T. F. Lei, D. Landheer, X. Wu, M. W. Ma, W. C. Wu, T. Y. Yang and T. S. Chao, in *Jpn. J. Appl. Phys.*, vol. 46, no. 10A, p. 6586, 2007.

19. W. Luo, Y. Kuo and W. Kuo, in *IEEE Trans. on Device and Materials Reliability*, vol. 4, no. 3, p. 488, 2004.
20. W. Luo, T. Yuan, Y. Kuo, J. Lu, J. Yan and W. Kuo, in *Appl. Phys. Lett.*, vol. 88, 202904, 2006.
21. W. Luo, T. Yuan, Y. Kuo, J. Lu, J. Yan and W. Kuo, in *Appl. Phys. Lett.*, vol. 89, 072901, 2006.
22. H. Satake and A. Toriumi, in *IEEE Trans. on Electron Devices*, vol. 47, no. 4, p. 741, 2000.
23. R. Degraeve, G. Groeseneken, R. Bellens, M. Depas and H. E. Maes, in *IEDM Tech. Dig.*, p. 863, 1995.
24. W. J. Zhu, T. P. Ma, T. Tamagawa, J. Kim, and Y. Di, in *IEEE Electron Device Letters*, vol. 23 no. 2, p. 97, 2002.
25. Z. Xu, M. Houssa, S. De Gendt and M. Heyns, *in Appl. Phys. Lett.*, vol. 80. no. 11, p. 1975, 2002.
26. T. Hori, *Gate dielectrics and MOS ULSIs—Principles, technologies and applications*, Berlin: Springer-Verlag, vol. 34, p. 44–45, 1997.
27. J. J. Lee, X. Wang, W. Bai, N. Lu and D. L. Kwong, in *IEEE Trans. on Electron Devices*, vol., no. 10, p. 2067, 2003.
28. Y. Kuo, J. Lu, S. Chatterjee, J. Yan, H. C. Kim, T. Yuan, W. Luo, J. Peterson, and M. Gardner, in *ECS Trans.*, vol. 1, no. 5, p. 447, 2006.
29. S. M. Sze, *Physics of Semiconductor Devices*, New York, John-Wiley & Sons, p. 403, 1981.
30. H. Cavendish and J. C. Maxwell, *Electrical Researches of the Hourable Henry Cavendish, F.R.S.*, University Press, p. 432, 1879.
31. R. Wan, J. Yan, Y. Kuo, and J. Lu, in *Jpn. J. Appl. Phys.*, vol. 47, no. 3, p. 1639, 2008.

Magnetic Resistive RAM

Mater. Res. Soc. Symp. Proc. Vol. 1160 © 2009 Materials Research Society   1160-H03-04-FF03-04

Novel Magnetoresistive Structures Using Self-Assembly and Nanowires on Si

Mazin Maqableh*, Xiaobo Huang and Bethanie J. H. Stadler
Department of Electrical and Computer Engineering, University of Minnesota, Minneapolis, Minnesota 55455

ABSTRACT

Anodic Aluminum Oxide (AAO) was grown both as free-standing membranes and as integrated layers on Si as templates for arrays of magnetoresistive nanowires. The barrier layer was completely removed in both cases and Co/Cu multilayered nanowires were successfully grown in these templates by DC electrodeposition. Magnetic hysteresis loops and current-perpendicular-to-plane giant magnetoresistance (CPP-GMR) up to 25% were measured for nanowires grown in free standing AAO templates and in templates grown on Si. Spin transfer torque (STT) switching was also measured for multilayers grown in free standing templates with a switching current density of 2.7 x 10^8 A/cm^2.

INTRODUCTION

Anodic Aluminum Oxide (AAO) is a promising template material for fabricating nanowires because of its self-assembled nanopores whose dimensions can be precisely controlled by tuning the different anodization parameters [1]. As well as having free standing AAO templates, the AAO can be integrated onto Si substrate [2-5] to open the road of making devices such as MRAM and catalysts that benefit from the combination of silicon processing and the self assembly properties of AAO.

AAO templates can be made using two-step anodization which results in highly ordered and straight nanopores [6]. A major concern for integrated nanowires is the removal of the barrier layer, which is a thin aluminum oxide layer existing at the bottom of the pores. This must be removed or thinned before efficient electrochemical deposition of nanowires can occur. Several methods have been used to remove this barrier. One method involves pore widening by phosphoric or chromic acid which will result in removal of the barrier layer as well as widening the pores [3, 5]. This method is disadvantageous in the sense that pore size is not preserved. Another method uses Ar ion-milling to break the barrier layer [7]. This method has two disadvantages. It requires an ultra thin AAO template so that Ar ions can reach the bottom of the pores with sufficient energy to break the barrier layer. It also damages the surface of the AAO as well as etching it, so the AAO thickness in this method is not preserved. A third method is to perform the second anodization for a very long time. A spike in the time dependence current curve during this step is used as a sign to manually stop the anodization process [3, 4].

Metallic nanowires can be grown in these templates by DC [3, 4, 8] and AC [9, 10] electrochemical deposition. Co/Cu multilayered nanowires have also been electrodeposited in free standing AAO templates using a mixture electrolyte that contains both Co and Cu cations [8, 12-14]. These electrochemically deposited multilayered nanowires have shown current-perpendicular-to-plane giant magnetoresistance (GMR) [8, 11] as well as a spin transfer torque phenomenon (STT) [12].

In this work, anodic aluminum oxide (AAO) was grown as free standing templates and also as successfully integrated templates on Si. The three barrier removal methods described above were tested here. Only the third method worked in the complete removal of the barrier layer which was further investigated by electrodepositing Cu into pores after attempting barrier removal. Therefore, this method was used in all the subsequent work presented in this paper. Co/Cu multilayered nanowires were successfully grown and their magnetic properties such as MH loops, giant magnetoresistance (GMR) have been measured. Spin transfer torque switching was also measured in the nanowires that were grown in free standing AAO templates.

EXPERIMENTAL

Free standing AAO templates

Two-step anodization [6] was used to make AAO templates from electropolished foils of Al metal. After anodization, the remaining Al metal was etched away with mercuric chloride, leaving oxide templates that contained nanopores with diamters 10 nm to 70 nm (using sulphuric acid, H_2SO_4), or from 40 nm to 150 nm (using oxalic acid, $H_2C_2O_4$). The barrier layers were etched by floating the templates on a mixture of phosphoric and chromic acid, and Cu films were sputtered onto the back of the templates.

AAO on Si

A 1µm aluminum film was evaporated using e-beam evaporation on a silicon substrate coated with titanium and copper films (200nm each). Two-step anodization process at 18C in 0.4M oxalic was then used to make a 600-700nm anodic aluminum oxide (AAO). The anodization voltage was kept constant at 40V during the two anodization steps. After the first anodization which was run for 4 minutes, the resulting AAO was etched away using a mixture of 1.8wt% chromic acid and 6wt% phosphoric acid for 30-45 minutes at 60C. The resulting aluminum, which was about 600nm thick, was second anodized using the same parameters to create a 600-700nm thick AAO with pore diameter of 40nm and inter-pore spacing of about 100nm. In addition to growing these latter pores directly onto Si, they were also grown onto Co(20nm)/Cu(10nm)/Co(10nm) thin films that were evaporated onto Si.

The barrier layer was removed by running the second anodization step for a much longer time. For the case of Si/Ti/Al substrate, which was used initially, this method failed in the removal of the barrier. However, for the case of Si/Ti/Cu/Al substrate, this method succeeded in removing this barrier.

Magnetorresistive nanowires.

DC electrochemical deposition was carried out at room temperature to grow Co/Cu multilayered nanowires in the membrane. The electrolyte solution was made of 155 g/L $CoSO_4.7H_2O$, 1.13 g/L $CuSO_4$ and 50g/L HBO_3. Cyclic voltametry was used to determine the cathode potential for Cu and Co deposition (-0.52 and -1 volts respectively) [13]. The purpose of HBO_3 was to maintain the pH value of the solution at 3.7.

Multilayered Co/Cu nanowires were fabricated with different layer thicknesses by controlling the deposition time of each layer. For the Si-integrated nanowires, 50 bi-layers of

36

Co(7.5nm)/Cu(5nm) were grown with a thick Cu layer (about 50nm) deposited prior to and after the deposition of the multilayered nanowires. Field emission scanning electron microscope (FESEM) was used to study the structure of the AAO and nanowires. Magnetic properties of the samples were verified by a vibrating sample magnetometer (VSM). MR was measured using an ac and dc magnetotransport systems with a bias current of 1mA. STT switching was measured in multilayered nanowires grown in the free standing AAO membranes.

RESULTS AND DISCUSSION

Free standing AAO templates

As the AAO was formed, using a two-step anodization process, columnar nanopores self-assembled inside the oxide to form a close-packed array. The pore diameters were varied from 10-60nm by changing the anodization conditions [1, 14]. As the diameter of the AAO nanopores decreased, the distance between the nanopores also decreased. The free-standing membranes had pores with lengths of 17µm.

AAO on Si

It was possible to vary nanopores diameter without changing spacing. This is a great way to analyze the affect of interwire magnetic interactions on the magnetoresistive (MR) properties, which will be done in the future. The initial MR characterization is reported. Figure 1 shows SEM images of self-assembled nanopores grown on Si with different pore sizes and fixed spacing.

Fig. 1. SEM images of self-assembled nanopores grown on Si (a) top view with pore size of 40nm and 100nm spacing (0.4M oxalic acid at room temperature), (b) top view with pore size of 30nm and 100nm spacing (0.3M oxalic acid at 5C), and (c) cross-sectional view of AAO on Si/Ti with a barrier layer.

Figure 2 shows the time dependence of the current during the long second anodization step. Fig. 2-a represents the second anodization of Al on Si/Ti substrate where the very long second anodization did not result in any spikes in the current, which means the barrier was not broken. Fig. 2-b represents the case of having Al on Si/Ti/Cu substrate where the current increased very rapidly at about 270s, and then came back to its low value at about 350s. The flat part on the top is from the instrument current limit which is 105mA. During this spike, the Cu film tended to be very reactive to the acid and it was completely gone after this spike. Thereafter, the sample behaved much likely as the Si/Ti/Al substrate. We thus used this spike as a sign to

37

manually stop the anodization process as shown in Fig. 2-c. Manually stopping the anodization process at this point of the curve was proven to be an ideal method to completely get rid of the barrier layer. However, if this anodization step was not given the sufficient time for the rapid increase in current to show up as shown in Fig. 2-d, the barrier layer remained at the bottom of the pores.

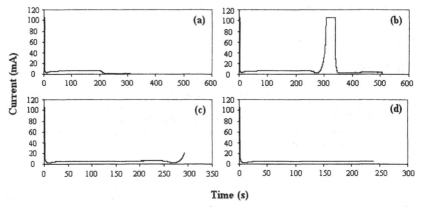

Fig. 2. Time dependence of the current during the (a) 2nd anodization of Al on Si/Ti substrate for 500s, (b) 2nd anodization of Al on Si/Ti/Cu substrate for 500s, (c) manually stopped 2nd anodization of Al on Si/Ti/Cu substrate at the early stages of the spike, and (d) manually stopped 2nd anodization of Al on Si/Ti/Cu substrate before the spike.

Attempts at electrochemical deposition inside the pores was used as evidence of the presence or the absence of a barrier layer in each sample. Figure 3 shows the behavior of the current during the electrodeposition of Cu in pores of samples whose barrier layers were treated using the three methods described above.

As seen in Fig. 3, only the third method resulted in a large, uniform current during deposition that can only be obtained when a good contact was made between the electrolyte and the metal electrode below the pores. The remaining barrier layer in samples processed by the first two methods essentially acts as an insulator that prevents the electrolyte from accessing the electrode.

Magnetoresistive nanowires

Using the free standing AAO templates, the highest magnetoresistance was found in nanowires that had hysteresis loops that were identical as measured in plane and perpendicular to the plane. The highest measured MR (= $\Delta R/R$ = 11%) of the multilayers was calculated as 33% by subtracting the resistance of the Cu leads on either side of the multilayers from the denominator [14]. Figure 4-a shows MH loops of a 50 bilayers of Co(7.5nm)/Cu(5nm), in which the sample appears to have an easy axis perpendicular to the nanowires.

It was found that multilayered Co/Cu nanowires grown in free standing AAO templates had similar magnetoresistance (MR) behavior as comparable nanowires grown on Si. However, compared to the relatively high MR value obtained in nanowires grown in free standing templates, nanowires on Si had a lower value which was measured to be 2-3%, Fig. 4-b,c. This

38

may be due to larger lead resistance, which is difficult to measure with these integrated samples, but further measurements are underway. The MR curve had a broader peak for the case where the field was applied parallel to the wires because the demagnetization fields due to the shape anisotropy of the Co layers inhibit switching until higher applied fields.

Spin transfer torque (STT) was measured in the nanowires that were inside free standing AAO. For 10-60nm diameter nanowires, the change in resistance due to STT was around 6% which represents the full magnetoresistance of the larger wires [14] but only half that of the smaller nanowires. It is therefore concluded that the 10-nm Co layers do not align antiparallel to parallel as fully at the switching current density of $J_{AP-P} = 2.7 \times 10^8$ A/cm^2 compared to the larger wires which switch at $J_{AP-P} = 3.2 \times 10^7$ A/cm^2. Au nanocontacts were electroplated into nanopores on Co(20nm)/Cu(10nm)/Co(10nm) thin films evaporated onto Si to prove the feasibility of using this technique in a wide range of configurations to study point-contact magnetoresistance and/or microwave response.

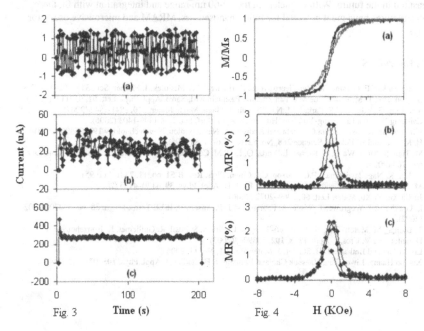

Fig. 3   Time (s)   Fig. 4   H (KOe)

Fig. 3. Time dependence of the current during DC electrodeposition of Cu in pores of samples whose barrier layers were treated using (a) Ar ion milling for 15mins, (b) pore widening in 5wt% H$_3$PO$_4$ for 45mins, and (c) very long second anodization.

Fig.4. MH loops of Co(7.5nm)/Cu(5nm) multilayered (50 bi-layers) nanowires with a diameter of 40nm when the field is perpendicular (blue) and parallel (pink) to the wires (a), and magnetoresistance when the field is perpendicular (b) and parallel (c) to the wires.

CONCLUSIONS

Anodic aluminum oxide (AAO) was successfully integrated as free standing membranes and as successfully integrated templates on Si with pore sizes that were varied between 10-60 nm by adjusting the anodization parameters. Into these templates, Co/Cu nanowires were successfully fabricated by DC electrodeposition with an easy axis perpendicular to the wires length. These mutlilayers showed a current-perpendicular-to-plane giant magnetoresistance (CPP-GMR) as well as spin transfer torque (STT) switching. Co/Cu multilayers grown on Si had a GMR ratio of 2-3% which is lower than that of the same multilayers grown in free standing templates. This might be due to the lead resistance in this integrated structure which is currently under study. Co/Cu multilayered nanowires grown in 10nm free standing AAO templates had a spin transfer torque (STT) phenomenon with a switching current density of 2.7 x 10^8 A/cm^2. Multilayers grown on Si are currently under construction and this switching phenomenon will be measured in the future. With diameters in the 10-60 nm range and integration with Si, these nanostructures have great potential for future nanosensors, MRAM and microwave oscillator arrays.

REFERENCES

1. A. Belwalkar, E. Grasing, W. Van Geertruyden, Z. Huang, W.Z. Misiolek, J. Membr. Sci., **319**, 192–198 (2008).
2. Bo Yan, Hoa T. M. Pham, Yue Ma, Yan Zhuang, Pasqualina M. Sarro, Appl. Phys. Lett. **91**, 053117 (2007).
3. S.K. Lim, G.H. Jeong, I.S. Park, S.M. Na, S.J. Suh, J. Magn. Magn. Mater. **310**, e841-e842 (2007).
4. Cai-Ling Xu, Hua Li, Guang-Yu Zhao, Hu-Lin Li, Appl. Surf. Sci. **253**, 1399-1403 (2006).
5. Filimon Zacharatos, Violetta Gianneta and Androula G Nassiopoulou, Nanotechnology **19**, 495306 (2008).
6. H. Masuda and K. Fukuda, Science **268**, No. 5216, 1466-1468 (1995).
7. M. Tian, S. Xu, J. Wang, N. Kumar, E.Wertz, Q. Li, P.M. Campbell, M. Chan, and T.E. Mallouk, *Nano Lett.* **5** , 697-703 (2005).
8. K. Liu, K. Nagodawithana, , P.C Searson, C.L Chien, Phys. Rev. B **51**, no. 11, 7381-4 (1995).
9. M. Kashi, A. Ramazani and A. Khayyatian, J. Phys. D: Appl. Phys. **39**, 4130–4135 (2006).
10. Jinxia Xu, Yi Xu, Mater. Lett. **60**, 2069–2072 (2006).
11. L. Gravier, J.-E Wegrowe, T. Wade, A. Fabian, and J.-P. Ansermet, IEEE Trans. Magn. **38**, no. 5, 2700-2702 (2002).
12. T. Blon,a_ M. Matefi-Tempfli, S. Matefi-Tempfli, L. Piraux, S. Fusil, R. Guillemet, K. Bouzehouane, C. Deranlot, and V. Cros, J. Appl. Phys. **102**, 103906 (2007).
13. Liwen Tan and Bethanie J. H. Stadler, J. Mater. Res. **21**, No. 11, 2006.
14. Xiaobo Huang, Liwen Tan, Haeseok Cho and Bethanie J. H. Stadler, J. Appl. Phys. **105**, 07D128 (2009).

**Poster Session:
Advanced Flash II**

Mater. Res. Soc. Symp. Proc. Vol. 1160 © 2009 Materials Research Society   1160-H04-01

Multi-Layered SiC Nanocrystals Embedded in SiO$_2$ Layer for Nonvolatile Memory Application

Dong Uk Lee[1], Tae Hee Lee[1], and Eun Kyu Kim[1*] Jin-Wook Shin[2] and Won-Ju Cho[2]

[1]Quantum-Function Spinics Laboratory and Department of Physics, Hanyang University, Seoul 133-791, Korea
[2]Department of Electronic Materials Engineering, Kwangwoon University, Seoul 139-701, Korea

ABSTRACT

A nonvolatile memory device with the multi-layered SiC nanocrystals embedded in the SiO$_2$ dielectrics for long-term data storage was fabricated and its electrical properties were evaluated. The multi-layered SiC nanocrystals were formed by using post thermal annealing process. The transmission electron microscope analysis showed that the multi-layered SiC nanocrystals are created between the tunnel and the control oxide layers. The average size and density of the SiC nanocrystals were approximately 5 nm and 2×10^{12} cm^{-2}, respectively. The memory window of nonvolatile memory devices with the multi-layered of SiC nanocrystals was about 2.7 V, and then it was maintained around 1.1 V after 10^5 sec.

INTRODUCTION

The trend in development of electronic memory devices is toward small size, high density, and fast programming/erasing speeds. Moreover, a nonvolatile memory based on the charge storage in discrete charge traps requires a long retention time combined with these features [1]. As a charge trapping layer the silicon-oxide-nitride-oxide-semiconductor (SONOS) devices have a better performance than conventional flash memory [2]. However, they have problems with scaling down the structures of these devices. Above all, it was leakage current to the substrate because the tunnel oxide layer becomes ultra thin. Thus it doesn't have a large on/off ratio for device performance. The nano-floating gate memory (NFGM) is a structure which has a few nano-sized particles isolated in dielectric materials and distributed between the tunnel and the control oxide layer. Due to these features, the NFGM has an advantage of preventing leakage current and improves the data retention and endurance. Recently, many studies have been reported about the NFGM including the various kinds of nanocrystals [3-10]. Especially, the SiC is a compound semiconductor belonging to IV group including SiGe. And it has many polytypes of different crystal structures with the cubic, the hexagonal or the rhombohedral, etc. The work function of SiC is about 4.0 ~ 4.5 eV. Also, its thermal stability is superior rather than metal materials. [11-14].

In this study the silicon-on-insulator (SOI) NFGM device structure with multi-layered SiC nanocrystals embedded in SiO$_2$ dielectrics was fabricated and its electrical properties were evaluated as a function of temperature such as subthreshold characteristics (V$_g$-I$_d$), output characteristics (V$_d$-I$_d$), threshold voltage shift, and retention times.

EXPERIMENT

The NFGM devices with SiC nanocrystals formed in the SiO_2 dielectric were fabricated on the p-type (100) UNIBOND SOI wafers. These wafers consist of a thickness of 200 nm buried oxide layer sandwiched between Si layers, and the depth of buried oxide layer is 100 nm. The tunnel oxide layer of 4.5-nm-thick was grown in a dry oxidation process. Subsequently, a SiC layers with a thickness of 10~15 nm were deposited on the tunnel oxide layer by using radio frequency magnetron sputtering. Also, an additional SiO_2 layer with a 50-nm-thick was deposited on a SiC layer by same sputtering system. The first post-annealing process was carried out at 900 °C for 3 min by using the rapid thermal annealing system in N_2 ambient. Through this annealing process, SiC nanocrystals were formed. Afterward, the control oxide layer of 30-nm-thick was deposited by magnetron sputtering system. And the second post-annealing process was carried out at 900 °C for 30 sec. The reason of second post-annealing process is a densification of control oxide layer grown by sputtering method. An aluminum layer on top side of the sample was deposited on the 150-nm-thick by thermal evaporation system and the gate electrode was formed by photolithography and Al etching process. Finally, the phosphorus plasma doping at 500 °C was carried out for the doping of source-drain region of the NFGM devices. The channel length and width were in the range of 2 ~ 20 μm [15,16].

DISCUSSION

Figure 1 (a) shows a cross-sectional TEM image of the NFGM with SiC nanocrystals embedded in SiO_2 dielectrics. After the first post-annealing process, the SiC nanocrystals had an irregular spherical shape, and their average size and density were estimated about 5 nm and $2x10^{12}$ cm^{-2}, respectively. These nanocrystals were distributed in the range of 30-nm-thick between the tunnel and the control oxide layer. Then, the SiC nanocrystals have a multi-layered structure mixed in additional SiO_2 layer. Figure 1 (b) illustrates a cross-sectional schematic diagram of the NFGM with the SiC nanocrystals embedded in the SiO_2 dielectrics. The SiC nanocrystals embedded in the SiO_2 dielectrics serve as the storage node. The channel length and width were in the range of 2 ~ 20 μm. The thicknesses of tunnel and control oxide layers were 4.5 nm and 30 nm, respectively.

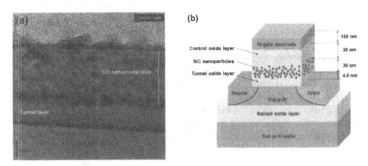

Figure 1 (a) Cross-sectional FE-TEM image of the SiC nanocrystals and (b) The cross-sectional schematic diagram of the nano-floating gate memory with the multi-layered SiC nanocrystals.

Figure 2 (a) shows that the capacitance-voltage (C-V) hysteresis curves of the nano-floating gate capacitor with the SiC nanocrystals at 25 °C. When the gate voltage sweeping was done in the ranges of ±8 V, ±10 V, ±12 V and ±14 V, the flat-band voltage shift (ΔV_{FB}) were about 2.2 V, 2.4 V, 3.2 V and 2.8 V, respectively, showing the counterclockwise C-V hysteresis curve. These C-V hystereses indicate the significant electron charging effect. The electron was injected in the SiC nanocrystals through tunneling oxide from p-Si substrate by using Fowler-Nordheim (FN) tunneling method. The C-V hysteresis curves in Fig. 2 (b) and (c) were measured at 85 °C and 125 °C, respectively. In these figures, the ΔV_{FB} appeared about 1.6 V at 85 °C and 0.6 V at 125 °C, when the gate voltages were swept from -8 V to 8 V. Also, the observed C-V hysteresis directions were counterclockwise and ΔV_{FB} were decreased about 27.8 % and 72.8 % at temperature of 85 °C and 125 °C, respectively. Figure 2 (d) shows the retention characteristics of the nano-floating gate capacitor with the SiC nano-particles at different temperatures of 25 °C and 85 °C. Here, the 700-ms-pulses of +12 V for programming operation and −12 V for the erasing operation were applied. Then, the ΔV_{FB} appeared approximately 0.54 V at 25 °C and 0.61 V at 85 °C after 1 hr.

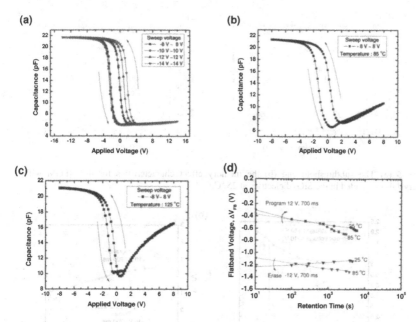

Figure 2 The C-V hysteresis curve of the nano-floating gate capacitor with SiC nanocrystals at (a) 25 °C, (b) 85 °C and (c) 125 °C. (d) The retention characteristics for the nano-floating gate capacitor with the SiC nanocrystals at 25 °C and 85 °C after the applied pulses with ±12 V for 700 ms.

The subthreshold characteristics of the NFGM with SiC nanocrystals embedded in SiO_2 dielectrics are shown in Fig. 3(a). When the drain voltages were induced from 0.05 to 1.00 V, the drain current was saturated in the range of 10^{-6} A. Also, the subthreshold slopes appeared 203 mV/dec. Then, the output drain current was about 3.2 μA/μm at $V_G - V_T = 2$ V and $V_{DS} = 1.0$ V. The effect of charging and discharging into the SiC nanocrystals were corroborated by the threshold voltage shifts after applied pulse bias. Figure 3(b) shows the subthreshold characteristics as a function of applying gate biases of +10 V and -10 V for 500 ms. As a results, in the programming state, the threshold voltage was shifted to 1.6 V toward a plus gate voltage as compared with the initial state. The electrons were injected from the channel of the NFGM to charge trap layer and quantum well of the SiC nanocrystals by FN tunneling. On the other hand, in the erasing state, the threshold voltage was shifted to 1.2 V toward a minus gate voltage as compared with the initial state. The confined electrons into quantum well or trap site around the SiC nanocrystals were emitted into the channel of NFGM device by FN tunneling. Therefore, the NFGM device with the SiC nanocrystals embedded in the SiO_2 dielectrics apparently has a memory effect with carrier charging.

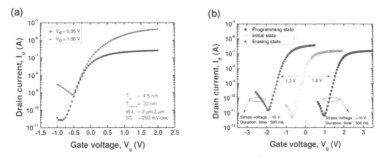

Figure 3 (a) The subthreshold and (b) the memory effect characteristics for the NFGM with SiC nanocrystals embedded in the SiO_2 dielectrics at 25 °C.

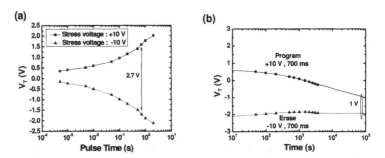

Figure 4 (a) The threshold voltage shifts of programming and erasing states as a function of applied pulse time. (b) The retention characteristics under the pulses of programming/erasing applied at +10 V and -10 V during 700 ms at 25 °C.

Figure 4 (a) shows the threshold voltage shifts of programming and erasing states as a function of applied pulse time. The pulses of the programming operation at +10 V and the erasing operation at -10 V were applied. Then, the memory window was observed at 2.7 V, when the saturation times of the threshold voltage for the programming and erasing states were at 700 ms. The retention characteristics of the NFGM with the multi-layered SiC nanocrystals embedded in the SiO_2 dielectrics are shown in Fig. 4(b). The initial memory window was 2.7 V, when the pulses applied for the programming/erasing at +10 V/-10 V during 700 ms. In the extrapolation to obtain predictable long-term data storage ability it was appeared about 1 V after 10^5 sec.

CONCLUSIONS

The NFGM device and the nano-floating gate capacitor with the multi-layered SiC nanocrystals embedded in the SiO_2 layer were fabricated and their electrical properties were evaluated at various temperatures. The morphology of the SiC nanocrystals showed a spherical shape with an average diameter of 3 ~ 5 nm. And a multi-layered structure was observed between the tunnel and the control oxide layers. The ΔV_{FB} of the nano-floating gate capacitor was about 1.6 V at 85 °C. The memory window of the NFGM with the multi-layered SiC nanocrystals was initially observed about 2.7 V at 700 ms, and it was maintained at about 1.1 V after 10^5 sec. These results indicate that the multi-layered SiC nanocrystals in the SiO_2 dielectrics have feasibility for the application in the next-generation nonvolatile memory devices.

ACKNOWLEDGMENTS

This work was supported in part by the National Program for 0.1-Terabit Non-volatile Memory Devices and the National Research Laboratory Program funded by the Korean Government.

REFERENCES

1. H. I. Hanafi, S. Tiwari, I. Khan, *IEEE Trans. Electron Devices*, **43**, 1553 (1996).
2. S. Tiwari, F. Rana, H. Hanafi, A. Hartstein, E. F. Crabbe, and K. Chan, *Appl. Phys. Lett.* **68**, 1377 (1996).
3. D. U. Lee, M. S. Lee, J.-H. Kim, E. K. Kim, H.-M. Koo, W.-J. Cho, and W. M. Kim, *Appl. Phys. Lett.* **90**, 093514 (2007).
4. M. S. Lee, D. U. Lee, J.-H. Kim, E. K. Kim, W. M. Kim, and W.-J. Cho, *Jpn. J. Appl. Phys.* **46**, 6202 (2007).
5. W.-R Chen, T.-C. Chang, P.-T. Liu, J.-L. Yeh, C.-H. Tu, J.-C. Lou, C.-F. Yeh, and C.-Y. Chang, *Appl. Phys. Lett.* **91**, 082103 (2007).
6. C.-W. Liu, C.-L. Cheng, S.-W. Huang, J.-T. Jeng, S.-H. Shiau, and B.-T. Dai, *Appl. Phys. Lett.* **91**, 042107 (2007).
7. H.-M. Koo, W.-J. Cho, D. U. Lee, S. P. Kim, and E. K. Kim, *Appl. Phys. Lett.* **91**, 043513 (2007).
8. J. Dufourcq, S. Bodnar, G. Gay, D. Lafond, P. Mur, G. Molas, J. P. Nieto, L. Vandroux, L. Jodin, F. Gustavo, and Th. Baron, *Appl. Phys. Lett.* **92**, 073102 (2008).

9. Y. Zhu, B. Li, and J. Liu, *J. Appl. Phys.* **101**, 063702 (2007).
10. M. She and T.-J. King, *IEEE Trans. Electron Devices* **50**, 1934 (2003).
11. Y. P. Guo, J. C. Zheng, A. T. S. Wee, C. H. A. Huan, K. Li, J. S. Pan, Z.C. Feng, and S. J. Chua, *Chem. Phys. Lett.* **339**, 319 (2001).
12. J.-J. Li, S.-L. Jia, X.-W. Du, and N.-Q. Zhao, *Surf. Coat. Technol.* **201**, 5408 (2007).
13. V. Kulikovsky, V. Vorlicek, P. Bohac, M. Stranyanek, R. Ctvrlik, A. Kurdyumov, and L. Jastrabik, *Surf. Coat. Technol.* **202**, 1738 (2008).
14. D. Song, E.-C. Cho, Y.-H. Cho, G. Conibeer, Y. Huang, S. Huang, and M. A. Green, *Thin Solid Films* **516**, 3824 (2008).
15. W.-J. Cho, C.-G. Ahn, K. Im, J.-H. Yang, J. Oh, I.-B. Baek, and S. Lee, *IEEE Electron Devices Lett.* **25**, 366 (2004).
16 T. H. Lee, D. U. Lee, S. P. Kim, and E. K. Kim, *Jpn. J. Appl. Phys.* **47**, 4992 (2008).

Mater. Res. Soc. Symp. Proc. Vol. 1160 © 2009 Materials Research Society   1160-H04-02

Fabrication and Electrical Characterization of Metal-Silicide Nanocrystals for Nano-Floating Gate Nonvolatile Memory

Seung Jong Han[1], Ki Bong Seo[1], Dong Uk Lee[1], Eun Kyu Kim [1*], Se-Mam Oh[2] and Won-Ju Cho[2]

[1]*Quantum-Function Spinics Laboratory and Department of Physics, Hanyang University, Seoul 133-791, Korea*
[2]*Department of Electronic Materials Engineering, Kwangwoon University, Seoul 139-701, Korea*

ABSTRACT

We have fabricated the nano-floating gate memory with the $TiSi_2$ and WSi_2 nanocrystals embedded in the dielectrics. The $TiSi_2$ and WSi_2 nanocrystals were created by using sputtering and rapidly thermal annealing system, and then their morphologies were investigated by transmission electron microscopy. These nanocrystals have a spherical shape with an average diameter of 2-5 nm. The electrical properties of the nano-floating gate memory with $TiSi_2$ and WSi_2 nanocrystals were characterized by capacitance-voltage (C-V) hysteresis curve, memory speed and retention. The flat-band voltage shifts of the $TiSi_2$ and WSi_2 nanocrystals capacitors obtained appeared up to 4.23 V and 4.37 V, respectively. Their flat-band voltage shifts were maintained up to 1.6 V and 1 V after 1 hr.

INTRODUCTION

Recently, the information technology has been developed rapidly and high technologies were occurred by digital convergence. The non-volatile memory devices occupy an important position in the memory market. However, the conventional non-volatile memory devices aim to achieve higher density, lower power consumption, and faster speed operation. To solve these improvements, non-volatile memory technologies such as nano-floating gate memory, ferroelectric memory, magnetoresistive memory, and phase change memory have been reported [1]. Especially, nano-floating gate memory (NFGM) based on nanocrystals is a very attractive candidate for future non-volatile memory application [2]. The NFGM devices effectively prevent the leakage current through the tunnel oxide layer and allow a thinner tunnel oxide, smaller operating voltage, better retention property, and faster program/erase speed [3~5]. The metal nanocrystals make the deep quantum wells between control and tunnel oxide layer due to the difference of work functions [6~9]. There are serious limitations on the fabrication of metal nano-floating gate memory associated with metal diffusion problem and deformation of metal nanocrystals during thermal processes. Therefore, the metal-silicide nanocrystals have been studied for the charge storage nodes of NFGM devices. The metal-silicide nanocrystals have a larger work function, higher density of states, smaller energy perturbation, and compatibility to current device processing [10~12].

In this study, the non-volatile memory devices with metal-silicide nanocrystals embedded in SiO_2 dielectric layer were fabricated by using sputtered metal-silicide thin film through post annealing. Also, the flat-band voltage shifts, the subthreshold, the threshold voltage shift, and the retention times were characterized.

EXPERIMENT

The nano-floating gate memory with the TiSi$_2$ and WSi$_2$ nanocrystals embedded in SiO$_2$ dielectrics were fabricated on p-type (100) silicon wafers and the p-type (100) UNIBOND silicon-on-insulator (SOI) wafers which have a 200-nm-thick buried oxide layer under a 100-nm-thick top Si layer. After the cleaning wafers, a tunnel oxide layer of 4.5-nm-thickness was grown by using dry oxidation process. Then, the metal-silicide layer with thickness ranges of 2 nm ~ 5 nm were deposited on the tunnel oxide layer by using direct current (DC) magnetron sputtering system. Subsequently, the WSi$_2$ nanocrystals are created on the tunnel oxide layer after rapid thermal annealing (RTA) under N$_2$ ambient at 1000 °C for 1 min. Then, the control oxide layer with 30 nm thicknesses was deposited by using radio frequency (RF) sputtering system. Also, the TiSi$_2$ nanocrystals were created between the contol and tunnel oxide layer after post-thermal annealing at 900 °C for 30 sec ~ 120 sec. To fabricate the metal-oxide-semiconductor structure, the aluminum gate electrode of 150-nm-thickness was evaporated by using thermal evaporator. Finally, the phosphorus plasma doping at 400 °C was carried out for the doping of source-drain regions of devices. [13] The channel length and width of the transistor are the ranges of 2 μm ~ 5 μm. The electrical properties of NFGM with the TiSi$_2$ and WSi$_2$ nanocrystals such as C-V hysteresis curve, flat-band voltage shifts, and retention characteristics were measured at room temperature by using pulse generator, precision LCR meter with a measurement frequency of 1-MHz and semiconductor parameter analyzer. The morphology of nanocrystals was analyzed by a field-emission transmission electron microscopy (FE-TEM).

DISCUSSION

Figure 1 shows the cross-sectional and the plane-view FE-TEM images of the TiSi$_2$ and WSi$_2$ nanocrystals embedded in the SiO$_2$ layer. The TiSi$_2$ and WSi$_2$ nanocrystals are spherical shape with an average diameter of approximately 2-5nm as shown in Fig. 1 (a) and (b).

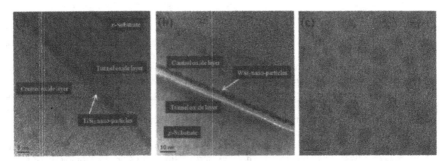

Figure 1 Cross-sectional FE-TEM image of (a) TiSi$_2$ and (b) WiSi$_2$ nanocrystals embedded in the SiO$_2$ layer. (c) The plane-view FE-TEM image of TiSi$_2$ nanocrystals.

Figure 2 (a) and (b) show the C-V characteristics of the nano-floating gate capacitors with TiSi$_2$ and WSi$_2$ nanocrystals. The flat-band voltage shifts appeared about 4.23 V and 4.37 V when gate voltage swept from -7 V to 7 V, respectively. Also, the observed C-V hysteresis direction is counterclockwise. This C-V hysteresis indicates a significant carrier charging effect; the carrier was injected in these nanocrystals through tunnel oxide layer from the p-Si substrate. As a result, the flat-band voltage shifts due to the charging effect were about 4.23 V and 4.37 V in a nano-floating gate capacitor with the $TiSi_2$ and WSi_2 nanocrystals. The flat-band voltage shifts of the nano-floating gate capacitors without $TiSi_2$ and WSi_2 nanocrystals were very small, the flat-band voltage shifts appeared about 0.2 V and 0.06 V under the gate voltage swept from -7 V to 7 V. This result implies nearly no carrier charging in this capacitor without nanocrystals. Therefore, the $TiSi_2$ and WSi_2 nanocrystals apparently have a memory effect.

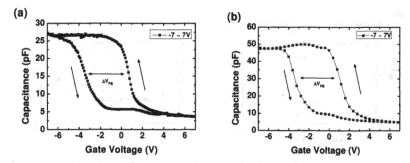

Figure 2 The C-V hysteresis curves of the nano-floating gate capacitors with (a) $TiSi_2$ and (b) WSi_2 nanocrystals.

Figure 3 (a) shows the program/erase speed of nano-floating gate capacitor with $TiSi_2$ nanocrystals. The stress time of the program/erase operation at +13 V / -13 V was applied during 1 s. Then, the flat-band voltage shift was observed about 3.5 V, when the saturation times of the flat-band voltage for program and erase operations were 20 ms. Also, as shown in Fig. 3 (b), the program/erase speed of nano-floating gate capacitor with WSi_2 nanocrystals were 10 ms at +7 V and 500 ms at -10 V. This result implies that the observed memory effect in fabricated nano-floating gate capacitor is charging of $TiSi_2$ and WSi_2 nanocrystals.

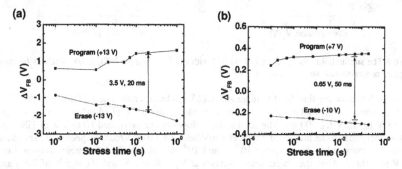

Figure 3 The program/erase speed of the nano-floating gate capacitor with (a) $TiSi_2$ and (b) WSi_2 nanocrystals.

Figure 4 shows the retention characteristics of the nano-floating gate capacitor with $TiSi_2$ and WSi_2 nanocrystals. The flat-band voltage shifts of the nano-floating gate capacitor with $TiSi_2$ nanocrystals decreased from 2.7 V to 1 V after 1 hr as shown in Fig. 4 (a), when the stress voltages for program/erase operations were at +13 V for 70 ms and at -12 V for 50 ms, respectively. After the stress voltages of program at +7 V for 10 ms and erase at -10 V for 500 ms were applied, the flat-band voltage shifts of the WSi_2 sample were decreased from 2.4 V to 1.6 V after 1 hr as shown in Fig 4 (b). This result implies that the stored charges in nanocrystals were leaked out. This charge loss might be originated from problem of fabrication process.

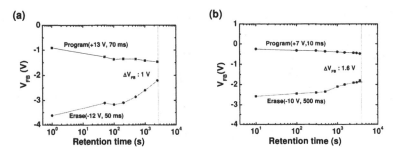

Figure 4 The retention time of the nano-floating gate capacitor with (a) $TiSi_2$ and (b) WSi_2 nanocrystals.

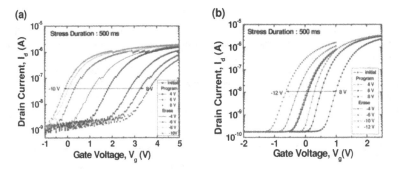

Figure 5 The subthrehold characteristics of NFGM with (a) $TiSi_2$ and (b) WSi_2 nanocrystals as function of applied program/erase voltage from +8 V to -12 V.

Figure 5 shows the threshold voltage shifts (ΔV_{th}) characteristics of NFGM with the $TiSi_2$ and WSi_2 nanocrystals including the tunnel layer as function of applied program/erase voltage from +8 V to -12 V for 500 ms. The subthreshold slopes of the fabricated NFGM with $TiSi_2$ and WSi_2 nanocrystals were measured about 165 mV/dec and 150 mV/dec, respectively. The drain current was saturated in the range of 10^{-4} A and 10^{-6} A under applied the drain voltages from 0.05 V to 1.00 V. Also, the output drain currents at $V_g - V_t = 2$ V and $V_{ds} = 3$ V of $TiSi_2$ and WSi_2 nanocrystals memory were about 45 μA and 20 μA, respectively. The ΔV_{th} of the NFGM with the $TiSi_2$ nanocrystals including the create tunnel layer of SiO_2/Si_3N_4 appeared about 4.1 V after applied program/erase bias at from 8 V to -10 V for 500 ms as shown in Fig 5 (a). Also, after applied the previous same condition, the ΔV_{th} of the NFGM with the WSi_2 nanocrystals including the tunnel layer of $Si_3N_4/SiO_2/Si_3N_4$ was approximately 1.7 V as shown in Fig 5 (b). In this structure, electrons were charged into the metal-silicide nanocrystals through the tunneling layer under the Fowler-Nordheim tunneling conditions.

CONCLUSIONS

The NFGM with the $TiSi_2$ and WSi_2 nanocrystals with average diameter of 2-5nm were fabricated between the control and the tunnel oxide layer during post-thermal annealing. Also, the electrical properties of NFGM devices such as the carrier charging effect, memory speed and retention were characterized. The flat-band voltage shifts of the $TiSi_2$ and WSi_2 nanocrystals capacitors were 4.23 V and 4.37 V during the gate voltage swept in the range of ±7 V, respectively, and their flat-band voltage shifts were maintained up to 1.6 V and 1 V after 1 hr. As a result, the $TiSi_2$ and WSi_2 nanocrystals are expected to apply in high-integrated nonvolatile memory devices.

ACKNOWLEDGMENTS

This work was supported in part by the National Program for 0.1-Terabit Non-volatile memory Devices and the National Research Laboratory Program funded by the Korean Government.

REFERENCES

1. R. Bez, and A. Pirovano, *Mater. Sci. Semicond. Process* **7** 349 (2004).
2. Y. Shin, H. G. Yang, J. Lv, L. Pu, R. Zhang, B. Shen, and Y. D. Zheng, *International Conference on Solid-State and Integrated Circuits Technology Proceedings*, pp. 881-884 (2004).
3. K. Yano, T. Ishii, T. Hashimoto, T. Kobayashi, F. Murai, and K. Seki, *IEEE Trans. Electron Devices* **41** 1628 (1994).
4. J. D. Blauwe, *IEEE Trans. Nanotechnol.* **1**, 72 (2002).
5. A. Kanjilal, J. Lundsgaard Hansen, P. Gaiduk, A. Nylandsted Larsen, N. Cherkashin, A. Claverie, P. Normand, E. Kapelanakis, D. Skarlatos, and D. Tsoukalas, *Appl. Phys. Lett.* **82**, 1212 (2003).
6. J. Dufourcq, S. Bodnar, G. Gay, D. Lafond, P. Molas, J. P. Nieto, L. Vandroux, L. jodin, F. Gustavo, and Th. Baron, *Appl. Phys. Lett.* **92**. 073102 (2008).
7. P. K. Singh, G. Bisht, R. Hofmann, K. Singh, N. Krishna, and S. Mahapatra, *IEEE Electron Device Lett.* **29**, 1389 (2008).
8. K. C. Chan, P. F. Lee, and J. Y. Dai, *Microelectron. Eng.* **85**, 2385 (2008).
9. B. Park, S. Choi, H–R. Lee, K. Cho. and S. Kim, *Solid State Commun.* **143**, 550 (2007).
10. S. M. Chang, H. Y. Huang, H. Y. Yang, and L. J. Chen, *Appl. Phys. Lett.* **74**, 224 (1999).
11. S. Bharat, P.K. Sahoo, and M. Katiyar, *Thin Solid Films* **462**, 127 (2004).
12. D. Panda, A. Dhar, and A, S. K. Ray, *IEEE Trans. Electron Devices* **55**, 2403 (2008).
13. D. U. Lee, M. S. Lee, J-H. Kim, E. K. Kim, H-M. Koo, W-J. Cho, and W. M. Kim, *Appl. Phys. Lett.* 90, 093514 (2007).

Mater. Res. Soc. Symp. Proc. Vol. 1160 © 2009 Materials Research Society   1160-H04-04

Charge Storage Properties of Nickel Silicide Nanocrystal Layer Embedded in Silicon Dioxide

Yoo-Sung Jang and Jong-Hwan Yoon

Department of Physics, College of Natural Sciences, Kangwon National University, Chuncheon, Gangwon-do 200-701, Republic of Korea

ABSTRACT

Memory properties of nickel silicide nanocrystal monolayers embedded in silicon dioxide have been investigated. The nanocrystal layers were produced by thermal annealing of a sandwich structure comprised of ultrathin Ni film (0.2 nm) sandwiched between two silicon-rich oxide ($SiO_{1.57}$) layers. Average diameter and areal density is about 2.9 nm and 1.3×10^{12} cm^{-2}, respectively. Capacitance-voltage (*C-V*) measurements are shown to have *C-V* characteristics suitable for nonvolatile memory applications, including large memory window (~ 10 V), long retention time (> 10^7 s), and excellent endurance (> 10^6 program/erase cycles).

INTRODUCTION

The demand for nonvolatile memory (NVM) devices with smaller size, faster operating speed and larger storage capacity is rapidly rising and as a consequence there is considerable research efforts devoted to realizing such devices. One approach is to use a floating-gate transistor where the floating-gate consists of nanocrystal (NC) charge traps. This method enables the tunneling oxide thickness of memory cells to be thinner, which makes NVM devices smaller and faster. On the other hand, multibit cell is another approach to enhance the storage capacity of NVM devices. Previous studies have demonstrated that multibit cells can be achieved by employing a multilayered NC floating gate [1, 2] or an extremely large memory window [3]. These can be realized by fabricating a well-defined NC monolayer with large charge storage capacity

Metal nanocrystals have received particular attention because they have additional advantages over those of semiconducting nanocrystals, namely, an enhancement in charge storage capacity and retention time [4-7]. These suggest that NVM devices satisfying the present demands might be realized by employing the floating gates based on metallic nanocrystal monolayers. Recently, we demonstrated a simple method that produced a well-defined nickel silicide (NiSi) NC monolayer, which has also physical properties similar to those of pure nickel, by thermal annealing of a sandwich structure comprised of an ultrathin Ni film sandwiched between two silicon-rich oxide layers [8]. In this work, we reports memory properties of NiSi NC monolayer fabricated by the same method as suggested in our previous work [8]. Despite thin tunnel oxide layer (~ 5 nm) the NC layer is shown to exhibit characteristic memory properties suitable for multibit NVM applications, including large memory window, long-term retention, and excellent endurance.

EXPERIMENTAL DETAILS

NiSi NC monolayers were formed by thermal annealing of a sandwich structure comprised of a thin Ni film sandwiched between two silicon-rich oxide ($SiO_{1.57}$) layers. $SiO_{1.57}$ films were deposited on (100) oriented p-type silicon wafers at a substrate temperature of 300 °C by plasma-enhanced chemical vapor deposition (PECVD) using fixed flow rates of SiH_4 and N_2O. Ultra-thin Ni layers were deposited using a conventional thermal evaporation method, with sandwich structures made by alternate deposition of $SiO_{1.57}$ and Ni layers. The details of a structure for memory properties are as tunnel SiO_2/$SiO_{1.57}$/Ni film/$SiO_{1.57}$/control SiO_2. The tunnel and control oxide thicknesses were 5 nm and 15 nm, respectively, and the thicknesses of $SiO_{1.57}$ and Ni layers were 3 nm and 0.2 nm, respectively. Nucleation and growth of Ni-based NCs was achieved by thermal annealing of the sandwich structures at 900 °C for 2 hrs in a quartz-tube furnace using high purity nitrogen gas (99.999 %) as an ambient. MOS capacitors were fabricated for memory property measurements by evaporating Al through a mask with circular holes of area 0.03 mm^2. The microstructure of NCs was investigated by transmission electron microscopy (TEM) using a JEOL JEM 2010 instrument operating at 200 kV. High-frequency (1 MHz) capacitance-voltage measurements were performed at 300 K using a Keithley 590 capacitance meter and a Keithley 230 programmable voltage source.

RESULTS AND DISCUSSION

Figure 1(a) shows the schematic structure of the sample employed for the formation of NiSi nanocrystal monolayer, which was prepared by depositing a SiO_2 layer of 5 nm, a $SiO_{1.57}$ of 3 nm, a Ni layer of 0.2 nm, a $SiO_{1.57}$ of 3 nm and a SiO_2 layer of 15 nm onto a p-type Si substrate.

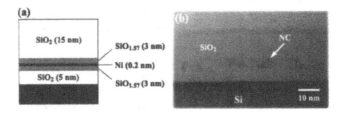

Fig. 1. (a) Schematic of the sample to produce Ni-based nanocrystal layer using a SiO_2/$SiO_{1.57}$/Ni /$SiO_{1.57}$/SiO_2 structure. (b) High-resolution cross-sectional TEM image of the sample.

56

Figure 1(b) shows high-resolution cross-sectional transmission electron microscopic (TEM) images of the sample after annealing at 900 °C for 2 hrs, respectively.

Fig. 1(b) clearly shows the formation of a well-defined NC monolayer hereby demonstrating the efficacy of the technique for making nanocrystal floating gate structures for nonvolatile memory devices. It is well known that Ni readily reacts with Si to form various Ni silicide phases, the exact phase depending on the annealing temperature [9]. Thermal annealing of a sandwich structure with ultrathin Ni film sandwiched between two $SiO_{1.57}$ layers, which contain excess Si atoms, is therefore likely to produce Ni silicide crystallites. In particular, our previous works [8, 10] showed that SiO_x layers with Ni atoms were transformed into SiO_2 forming NiSi NCs embedded in SiO_2 matrix. The size distribution of the NCs, as analyzed by using plane-view TEM micrographs, ranges from 1.8 nm to 4.0 nm, with an average diameter of 2.9 nm, and have an areal density of 1.3×10^{12} cm^{-2}.

The programming (P) and erasing (E) characteristics of the NC layer were performed using Fowler-Nordheim tunneling by applying a voltage on the gate electrode of the MOS devices. Figure 2 shows high-frequency (1 MHz) C-V hysteresis measured for the MOS capacitors with [Fig. 2(a)] and without [Fig. 2(b)] the NiSi NCs. [The data shown in Fig. 2(a) were obtained from the sample shown in Fig. 1, while Fig. 2(b) represents the results reported in our previous (ref. 8) work for comparison]. The sample structure in Fig. 2(b) was formed with the following layers: tunnel SiO_2 /$SiO_{1.57}$/control SiO_2. The tunnel and control oxide thickness were 5 and 7 nm, respectively, and the thickness of $SiO_{1.57}$ was 7 nm. The gate voltage was swept from negative to positive values before being swept back from positive to negative values. As seen in Fig.2, there is clearly a distinct difference in the memory window for the two cases. For MOS structures with NiSi NCs [Fig. 2(a)] the memory window increases, eventually followed by a saturation value of about 10 V, as the sweep range of the gate voltages increases. While MOS structures without

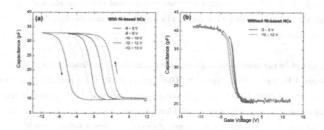

Fig. 2. C-V characteristics of MOS structures provided with the samples: (a) with NiSi nanocrystals and (b) without NiSi nanocrystals (ref. 8).

NiSi NCs [Fig. 2(b)] show no significant change in the memory window. This difference demonstrates that the large memory window shown in Fig. 2(a) is closely associated with the presence of the NiSi NCs. During applying positive voltages, the electrons directly tunnel from the Si substrate through the tunnel oxide, and are trapped at either NCs, or defects within SiO_2 or defects at the NC/SiO_2 interface. While applying negative voltages, the holes tunnel from the Si substrate and recombine with the electrons trapped at either the NCs or defects. However, comparison between the two cases demonstrates that the large memory window shown in Fig. 2(a) results from the electrons trapped at the NCs rather than the defects. In this case, the saturation of the memory window may be due to the Coulomb's repulsion. As estimated by using $Q=\Delta V.C_{ox}$, where Q is the total charge trapped at the NCs, ΔV is the saturated memory window of about 10 V, and C_{ox} is the oxide capacitance of 33 pF, the magnitude of the Q is 3.3×10^{-10} C. This value means that the number of electron trapped per NC is about 5.3, resulting in the large memory window. The large memory window supports that the NiSi nanocrystal layer produced in this study has great potential for multibit nonvolatile memory applications.

Figure 3 shows the charge retention characteristics after various P/E operations at room temperature.

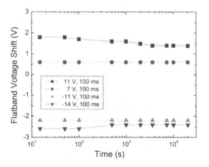

Fig. 3. Retention characteristics of MOS structures provided with the sample shown in Fig. 1. The retention time was about 10^7 s for the state programmed with a pulse of 11 V, 100 ms.

In spite of the thin tunnel oxide (< 5 nm), no significant shift in the flatband voltage (V_{FB}) is observed. For the low P/E voltages (7 V, 100 ms and -11 V, 100 ms), in particular, there is no V_{FB} shift up to 2×10^4 s, but for the higher voltage programmed state (11 V, 100 ms) one finds that the retention time, defined as the time when half of the stored charge is lost, is over 10^7 s. NiSi has been shown to have a work function of about 4.7 eV similar to that of pure Ni. As a result, it is believed that the excellent retention characteristics shown in Fig. 3 result from the large work function difference between the NiSi NC and Si substrate and thus, the suppression of the electron tunnel back to the substrate. The improved retention characteristics of V_{FB} shift supports the premise that NiSi NC layer is suitable for the memory devices with thin tunnel oxide layer, which is required to realize NVM devices with smaller size and faster operating speed.

Figure 4 shows the endurance characteristics of the sample shown in Fig. 1. Pulses of (+14 V, 30 ms) and (-15 V, 30 ms), which cause almost the fully programmed and fully erased states, were used to examine endurance characteristics for the P/E operations, respectively. The results clearly show there are no significant changes in the programmed and erased V_{FB} shifts up to 10^6 cycles except for a small increase in the programmed V_{FB} shift after 10^4 cycles.

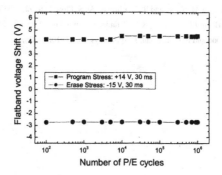

Fig. 4. Endurance characteristics of MOS structures provided with the sample shown in Fig. 1. Voltage pulses of (+14 V, 30 ms) and (-15 V, 30 ms) were applied to the gate electrode for P/E operations.

In conclusion, memory properties of NiSi nanocrystal monolayers produced by thermal annealing of a simple sandwich structure consisting of an ultra-thin metal (Ni) layer sandwiched between two silicon-rich oxide layers have been investigated. The capacitance-voltage characteristics of a structure based on a 0.2 nm Ni film sandwiched between two $SiO_{1.57}$ layers (3 nm) were shown to exhibit large memory window, long-term charge retention, and excellent endurance characteristics suitable for realizing multibit and small size nonvolatile memory applications.

ACKNOWLEDGMENT

This work was supported by the Korea Research Foundation Grant funded by the Korean Government (MOEHRD, KRF-2008-313-C00303).

REFERENCES

1.   T. Z. Lu, M. Alexe, R. Scholz, and M. Zacharias, Appl. Phys. Lett. **87**, 202110 (2005).
2.   J. Wang, L. Wu, K. Chen, L. Yu, X. Wang, J. Song, and X. Huang, J. Appl. Phys. **101**, 014325 (2007).
3.   S. Park, Y. K. Cha, D. Cha, Y. Park, I. K. Yoo, J. H. Lee, and K. S. Seol, Appl. Phys. Lett. **89**, 033122 (2006).

4. Z. Liu, C. Lee, V. Narayanan, C. Pei, and E. C. Kan, IEEE Trans. Electron Devices **49**, 1606 (2002).

5. Z. Liu, C. Lee, V. Narayanan, C. Pei, and E. C. Kan, IEEE Trans. Electron Devices **49**, 1614 (2002).

6. C. Lee, J. Meteer, V. Narayanan, and E. Kan, J. Electron Mater **34**, 1 (2005).

7. D. Zhao Y. Zhu, and J. Liu, Solid State Electronics **50**, 268 (2006).

8. Y. S. Jang, J. H. Yoon, and R. G. Elliman, Appl. Phys. Lett. **92**, 253108 (2008).

9. C. Hayzlden, J.L. Batstone, R.C. Cammarata, Appl. Phys. Lett. **60**, 225 (1992).

10. J. H. Yoon and R. G. Elliman, J. Appl. Phys. **99**, 116106 (2006).

Mater. Res. Soc. Symp. Proc. Vol. 1160 © 2009 Materials Research Society   1160-H04-05

Localized Silicon Nanocrystals Fabricated by Stencil Masked Low Energy Ion Implantation: Effect of the Stencil Aperture Size on the Implanted Dose

R. Diaz[1], C. Dumas[3], J. Grisolia[1], T. Ondarçuhu[2], S. Schamm[2], A. Arbouet[2], V. Paillard[2], G. BenAssayag[2], P. Normand[4], J. Brugger[5]

[1]INSA/ Univ. Toulouse, 135 Avenue de Rangueil, 31077 Toulouse, France
[2]CNRS-CEMES and Univ. Toulouse, 29 rue Jeanne Marvig, 31055 Toulouse Cedex 4, France
[3]IM2NP - Universités Paul Cézanne, Provence et Sud Toulon Var - France
[4]Institute of Microelectronics, NCSR 'Demokritos', 15310 Aghia Paraskevi, Greece
[5]EPFL Laboratoire de Microsystèmes, CH-1015 Lausanne, Switzerland

ABSTRACT

In this paper, we develop a new method based on ultra-low-energy ion implantation through a stencil mask to locally fabricate Si nanocrystals in an ultrathin silica layer. We perform a 1 keV Si implantation with doses of 5×10^{15} Si$^+$/cm^2, 7.5×10^{15} Si$^+$/cm^2 and 1×10^{16} Si$^+$/cm^2 in a 7 nm thick silicon oxide layer through stencil mask apertures ranging from 1μm up to 5 μm. After the mask removal the samples are furnace annealed at a temperature of 1050°C for 90 min under N$_2$ atmosphere. The samples are then characterized by mapping the implanted and non-implanted areas by atomic force microscopy and photoluminescence spectroscopy. The intensity and the wavelength of the PL peak are found to depend on the implanted NCs cell size. A slight blue shift from 730 nm up to 720 nm is observed with decreasing cell size. Simultaneously, the PL intensity decreases and the signal vanishes for submicron features (which should contain 10^2 to 10^3 NCs). AFM microcopy performed on the implanted regions shows that the well-known oxide swelling usually observed after NCs synthesis decreases from 3.5 nm down to 0 as the cell size decreases. This result demonstrates that the effective implanted dose clearly decreases with the size of the cells. This effect is probably due to an electrostatic charging of the Si$_3$N$_4$ membrane despite the metallization treatments applied to the mask surface.

INTRODUCTION

Silicon nanoparticles are of great interest for high density data storage applications. They were first introduced with the aim of replacing the poly-silicon floating gate of conventional flash memories with a layer of mutually isolated NPs [2,3,4]. This improves the immunity of the stored data against charge loss through weak spots in the oxide and thus, enables thinner tunnel oxides and further device downscaling. Furthermore, silicon NPs with a size of less than 10 nm have demonstrated single electron charging effects at room temperature [5,6] and appeared as attractive candidates for single electron memories in which the addition or subtraction of one electron can represent a bit of data [7,8].
To achieve reliable devices and avoid fluctuations in device performance not only the NCs characteristics (size, density, interspacing) but also their location inside the gate dielectric have to be well controlled. In particular, a fine control of the thickness of the dielectric layer (injection oxide) separating the NCs and the substrate is absolutely required since a fluctuation of less than 1nm affects dramatically the programming properties (write/erase times and voltages) and data retention of the devices. Among the different technological routes [2,9-11] explored the last years for generating NCs in the gate oxide of MOS devices, the ion-beam-synthesis (IBS) technique

has received substantial attention due to its flexibility and manufacturing advantages. The potential of this technique for NC-based-memories operating at low voltages has been enhanced through the synthesis in the ultra-low-energy (ULE) regime (typically 1keV) of single Si-NCs layers in thin SiO_2 films ($\leq$10nm) [1,9,12]. In this paper the method is extended by using a silicon nitride stencil mask (SM) during implantation to define and control the implanted zones. Atomic Force Microscopy (AFM) and Photoluminescence (PL) spectroscopy are used to characterize the Si NCs cells that form after thermal annealing. The main advantages of the SM-ULE-IBS technique are a full compatibility with the mainstream CMOS technology and a process flexibility allowing the generation of NPs with well-controlled size, depth distribution and wafer surface location.

EXPERIMENTAL RESULTS

Experimental details

The fabrication starts by implanting silicon ions into 7 nm thick thermally-grown oxide on top of an 8-in., p-type, (100)-oriented Si wafer. This implantation is realized at 1keV energy through a silicon nitride stencil mask having apertures ranging from 50 nm up to 5 μm. The stencil apertures were fabricated by focused-ion-beam or deep UV (DUV) lithography / etching processing. An aluminum, silver or gold thin layer was evaporated on the stencil mask to reduce any charging effects during ion implantation and to enhance the mask strength. The implantation doses are 5×10^{15}, 7.5×10^{15} and 1×10^{16} Si^+/cm^2. To restore the quality of the SiO_2 layer and generate Si-NCs, a post-implantation annealing treatment in N_2 atmosphere for 90 min at 1050°C is performed. The structural and optical studies reported herein relate to Si-implants performed through stencil apertures of 1.5 μm, 2 μm and 4 μm. Scanning electron microscopy (SEM) technique is used to image and to compare stencil apertures and implanted areas dimensions. PL spectra were recorded using a Dilor XY spectrometer at room temperature equipped with a LN2-cooled CCD, notch filter and a 150 groves/mm grating. The excitation wavelength is 488 nm using a power of 150 μW on the sample. A x100 and 0.95 NA objective is used to focus the incident laser light on the sample and to collect the light (backscattering). AFM microscopy in contact mode is also performed to image the implanted pockets and to measure the swelling values.

The SM-ULE-IBS technique

Figure 1b shows a SEM image of a sample implanted (1×10^{16} Si^+/cm^2) through a SM with oval apertures and subsequent thermally annealed. The implanted areas (dark areas on the SEM image) are identical to the mask apertures within a relative error lower than 10%. This result shows that the patterns are well transferred without the common blurring effect usually observed with stencil masks. It also points out that SM-ULE-IBS is an efficient method to generate Si-NCs in well-localized oxide cells with size in the μm^2 range.

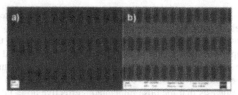

Fig.1: SEM image a) of the silicon nitride membrane including micrometric oval apertures (1.5μm x 5μm) realized by DUV lithography, and b) implanted areas (1.6μm x 4.9μm) of a sample realized by SM-ULE-IBS after implantation through the stencil mask and annealing.

Effect of the stencil mask aperture size on the swelling

AFM measurements of the oxide swelling as it appears after implantation and annealing, was carried-out for evaluating the effect of the SM aperture size on the implanted dose. Such measurements were first performed on a set of 3 samples implanted through the same stencil mask with an aperture of 2 μm um in diameter, to doses of 1×10^{16}, 7.5×10^{15} and 5×10^{15} Si$^+$/cm^2, and further annealed at 1050 °C for 90 min. As shown in figure 2, for a same aperture size, the swelling height increases linearly with the implanted dose going from about 1 nm (low dose) to 3 nm (high dose).

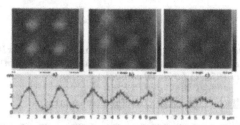

Fig. 2: AFM images and oxide swelling profiles of SM-ULE-IBS samples implanted to a dose of a) 1×10^{16} Si$^+$/cm^2, b) 7.5×10^{15} Si$^+$/cm^2, c) 5×10^{15} Si$^+$/cm^2 and subsequently thermally annealed.

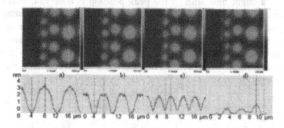

Fig. 3: AFM images and oxide swelling (OS) profiles of implanted areas with different diameters (D); a) D = 4.5μm, OS peak ≅ 3.5nm, b) D = 3.2μm, OS peak≅ 2.5nm, c) D = 2μm, peak ≅ 2.2nm, d) D = 1.5μm, OS peak ≅ 1.4nm

Figure 3 shows AFM images and profiles from the same sample implanted to a dose of $1x10^{16}$ Si^+/cm^2 through a stencil mask with apertures ranging from 1µm up to 5µm. As depicted in Figure 2, the oxide swelling of the implanted areas is uniform and the threshold between the implanted and non-implanted areas is abrupt. A systematic and unexpected "halo" appears on the edge surrounding the implanted areas with a height in the 0.1 nm range. Surprisingly, the maximum swelling measured in the center of these areas decreases with the aperture size (see Figure 4). This result suggests that the dose implanted ("effective dose") decreases as the stencil aperture size decreases. For the smallest sizes this "effective dose" is below the threshold of detection of NCs for an implantation dose of $5x10^{15}$ Si^+/cm^2. For the smallest sizes the oxide swelling is comparable to that detected for an oxide implanted to $5x10^{15}$ Si^+/cm^2 thus indicating a dose loss of about half of the programmed one ($1x10^{16}$ Si^+/cm^2).

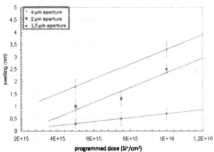

Fig. 4: evolution of the oxide swelling for 3 implantation dose values ($5x10^{15}$ Si^+/cm^2, $7.5x10^{15}$ Si^+/cm^2 and $1x10^{16}$ Si^+/cm^2) for 3 aperture sizes (1.5µm, 2µm and 4µm) after an annealing under N_2 atmosphere at 1050 °C during 90 minutes.

Stencil window size effect on the PL spectra of implanted areas

PL spectra on the Figure 5 are obtained in the center of micrometric patterns implanted to a dose of $5x10^{15}$ Si^+/cm^2. First, it is important to notice that the spectrum taken in the 5µm aperture is similar to a PL spectrum recorded in a non-masked area (~735nm). While decreasing stencil aperture dimension, a slight blue shift of the peak position is observed, starting from 735 nm position (5µm aperture) down to 715 nm (2µm aperture). This shift is associated to a decrease of the PL band intensity. Such a behavior can be attributed to the decrease of the nanocrystal average size. The above results indicate that the effective implanted dose is equal to the programmed one only for stencil apertures larger than 5 µm and decreases with the aperture size. For the lower programmed dose ($5x10^{15}$ Si^+/cm^2) and small stencil apertures, the effective dose is too low to form NCs.

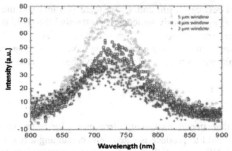

Fig.5: *PL spectra taken for different stencil aperture sizes: 2μm (blue curve), 4μm (black curve) and 5 μm (pink curve) after Si⁺ implantation (at 5x10¹⁵ Si⁺/cm² dose) and annealing under N₂ ambient at 1050°C during 90 minutes.*

DISCUSSION

We have shown that the intensity and the position of the PL peak decrease with the implanted pattern size. The decrease of the PL intensity leads to the disappearance of the signal for submicron features (which should contain 10^2 to 10^3 NCs). These results could derive either from an additional oxidation of the Si in excess during the 1050°C annealing, or from an effective implantation dose different from the expected value, the so-called "programmed" dose.

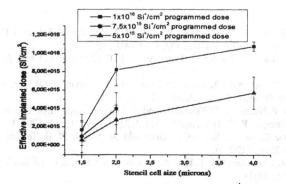

Fig.6 : *« Effective » dose implanted through 3 stencil mask windows of 1.5 μm, 2 μm and 4 μm and 3 programmed doses (1x10¹⁶ Si⁺/cm², 7.5x10¹⁵ Si⁺/cm² and 5x10¹⁵ Si⁺/cm²)*

The swelling values obtained from AFM microscopy which decrease from 3.5nm down to 0 with the decreasing size of the Si cells demonstrate that the effective implanted dose clearly decreases with the size of the features. Figure 6 summarizes these results (decrease of the implanted dose with the aperture size) for different doses (5x10¹⁵ Si⁺/cm², 7.5x10¹⁵ Si⁺/cm² and 1x10¹⁶ Si⁺/cm²) and stencil aperture sizes (1.5 μm, 2 μm and 4 μm). For the higher programmed dose (1x10¹⁶ Si⁺/cm²) and for aperture sizes larger than 5μm, the effective dose is similar to the programmed dose. For aperture sizes between 2 µm and 4 µm, there is a loss and the effective dose is still sufficient to obtain NCs, as previously noticed. The effective threshold dose of 5×10^{15} Si^+/cm^2 corresponds to aperture sizes between 1.5 µm and 2µm. Then, for submicron apertures the effective dose would remain at too low figures to obtain NCs. For programmed doses of 7.5 and 5×10^{15} Si^+/cm^2, the minimum aperture size for Si-NC synthesis shifts to around 2µm and 3.5 µm, respectively.

CONCLUSION

We have demonstrated that stencil masked ultra low energy ion beam synthesis can be a parallel, low cost and reliable method to produce isolated cells containing a few to several thousands of NCs. The number of Si-NCs is controlled by the mask geometry (aperture size), implantation dose and annealing conditions. We have also shown a surprising "dose effect" in terms of decrease of the implanted dose with the mask aperture size. Despite the single or double face metallization of the silicon nitride stencil mask, this dose loss might be due to some residual charging during ion implantation. The effective dose was found to decrease linearly with the aperture diameters. Nevertheless NCs were detected by PL on areas of 300x300 nm^2. With a programmed dose as high as 1×10^{16} Si^+/cm^2, the threshold effective dose (5×10^{15} Si^+/cm^2) necessary for NCs synthesis cannot be reached using a stencil mask with deep submicron apertures. At the moment the variation of the implanted dose with the aperture size is the main limitation of this method. Additional experiments using a metallic mask with submicron features would provide more insights on the stencil mask charging effect during ion implantation.

References

[1] C. Dumas, J. Grisolia, G. BenAssayag, V. Paillard, A. Arbouet, S. Schamm, P. Normand and J. Brugger, Phys. Stat. Sol. (a) 204, No.2, 487-491 (2007).
[2] S. Tiwari, F. Rana, H. I. Hanafi, A. Hartstein, E. F. Crabbé, and K. Chan, Appl. Phys. Lett. 68, 1377 (1996).
[3] H. I. Hanafi, S. Tiwari, and I. Khan, IEEE Trans. Electron Devices ED43, 1553 (1996).
[4] J. De Blauwe, IEEE Trans. Nanotechnol. 1, 72 (2002).
[5] E. Kapetanakis, P. Normand, D. Tsoukalas and K. Beltsios, Appl. Phys. Lett. 80, 2794 (2002).
[6] S. Huang, S. Banerjee, R. T. Tung, and S. Oda, J. Appl. Phys. 93, 576 (2003).
[7] G. Molas, B. De Salvo, D. Mariolle, G. Ghibaudo, A. Toffoli, N. Buffet and S. Deleonibus, Surf. Sci. Spectra 47, 1645 (2003).
[8] P. W. Li, W. M. Liao, David M. T. Kuo, S. W. Lin, P. S. Chen, S. C. Lu and M.-J. Tsai, Appl. Phys. Lett. 85, 1532 (2004).
[9] Normand P, Kapetanakis E, Dimitrakis P, Skarlatos D, Beltsios K, Tsoukalas D, et al. Nucl Instrum Methods Phys Res B 2004 ; 216 :228.
[10] Ammendola G, Vulpio M, Bileci M, Nastasi N, Gerardi C, Renna G, et al. J Vac Sci Technol B 2002 ; 20 :2075
[11] King YC, King TJ, Hu C. IEEE Trans Electron Dev Meet 2001; ED-48:696.
[12] Carrada M, Cherkashin N, Bonafos C, Ben Assayag G, Chassaing D, Normand P, et al. Mater Sci Eng B 2003 ; 101 :204.
[13] C.Bonafos, M.Carrada, N.Cherkashin, H.Coffin, D.Chassaing, G.BenAssayag and A.Claverie, T.Müller and K.H.Heinig, M.Perego and M.Fanciulli, P.Dimitrakis and P.Normand, Journal of Applied Physics, vol.95, n°10, p.5696, 2004.

Poster Session:
Charge Trap II and MRAM

Mater. Res. Soc. Symp. Proc. Vol. 1160 © 2009 Materials Research Society   1160-H05-08

HfO2-Based Thin Films Deposited by RF Magnetron Sputtering

L. Khomenkova[1], C. Dufour[1], P.-E. Coulon[2], C. Bonafos[2], F. Gourbilleau[1]
[1]CIMAP, UMR CNRS 6252, ENSICAEN, 6 B^d M^{al} Juin, 14050 Caen Cedex 04, France
[2]CEMES/CNRS, 29 rue J. Marvig, 31055 Toulouse Cedex 04, France

ABSTRACT

HfO2-based layers prepared by RF magnetron sputtering were studied by X-ray diffraction, infrared absorption spectroscopy and transmission electron microscopy techniques. Since the stability of amorphous structure of HfO2-based layers is interesting from microelectronic application point of view, the effect of the deposition parameters and post-deposition annealing treatment on the properties of the layers was investigated. The amorphous-crystalline transformation of pure HfO2 layers is observed to be stimulated by annealing treatment at 800 °C. It was found that the incorporation of silicon in HfO2 matrix allows to prevent crystallization of the layers and to shift the crystallization temperature to values up to 900 °C.

INTRODUCTION

Transition-metal-oxides have numerous applications in optics, chemistry, bio- and microelectronics. The latter domain has to face the continuing scaling down of CMOS devices linked with an increase of their number on industrial chips. Unfortunately, SiO2 and related oxynitrides met their fundamental limits as conventional gate dielectrics. Therefore, high-k gate dielectrics were investigated as alternative gate dielectrics. One of the promising materials is hafnium oxide [1,2]. It offers a high thermal stability, a wide band gap and a high permittivity which is the most important point required to circumvent the problem of high leakage current. However, in a thin film approach, the structure and the properties of HfO2 layers strongly depend on the deposition conditions and the post annealing treatment.

Various deposition techniques such as ALD [3-5], PVD [6], CVD [7] have been used for the fabrication of HfO2 films, though RF magnetron sputtering is not commonly addressed. Up to now the deposition of HfO2 layers has been performed by reactive sputtering of a pure metal Hf cathode in Ar+O2 gas mixture [8]. Only in a few cases, a pure HfO2 target was used for films fabrication [9,10]. A serious problem for the application of HfO2 films is its low crystallization temperature (much less than 500 °C) since, in the crystalline phase, the grain boundaries act as diffusion paths for oxygen or dopants into the dielectric gate. Thus, a lot of efforts were devoted to improve thermal stability of the amorphous structure. It was shown that nitrogen incorporation into the oxide matrix during deposition or post-deposition processes leads to the formation of HfON [8] and shifts the crystallization temperature up to 1000 °C [11]. It was also observed that amorphous HfSiON layers are stable after such annealing treatment, while the HfSiO ones are crystallised. In the two last cases the layers were grown by reactive sputtering of HfSi target in Ar+O2 and/or Ar+N gas mixture [11].

In the present study, the properties of pure and Si-rich HfO2 layers, fabricated by RF magnetron sputtering of pure or Si-topped HfO2 target, were studied versus deposition conditions and annealing treatment.

EXPERIMENT

The layers investigated were grown on Si substrates (B-doped, ~15 Ωcm, (100)) by RF magnetron sputtering of 4 inches pure HfO_2 target in pure Ar plasma. Before deposition, the substrates were cleaned in a diluted hydrofluoric solution (10%) to remove native oxide on the surface, leaving an hydrogen-terminated surface. After cleaning, they were immediately placed into the vacuum chamber of deposition setup.

The layers were grown at different RF powers (RFP=40-80 W) and substrate temperatures (T_S=45-500°C). The deposition was performed under total plasma pressures in the range P_{total}=0.027-0.110 mbar. Si-rich HfO_2 layers were grown by co-sputtering a pure HfO_2 target topped by Si chips. The Si incorporation was monitored through the surface covered by the Si chips which varies from 3 to 12% of the total target surface. The higher Si content in the layers was obtained with the highest Si surface ratio, R_{Si}, equal to 12%. An annealing treatment in the furnace under nitrogen flow at different temperatures (T_A=700-900°C) and durations (t_A=1-15 min) was performed. The thickness of the layers investigated here did not exceed 12 nm.

The properties of the layers were analyzed by means of X-ray reflection and X-ray diffraction (XRD), attenuated total reflectance infrared spectroscopy (ATR) and transmission electron microscopy (TEM). ATR spectra were measured in the range 600-4000 cm^{-1} by means of a 60° Ge Smart Ark accessory inserted in a Nicolet Nexus spectrometer. X-ray diffraction analysis was performed using a Phillips XPERT HPD Pro device with Cu K_α radiation (λ=0.154 nm) at a fixed grazing angle incidence of 0.5°. Cross-sectional (XS) specimens were prepared for TEM examination by the standard procedure involving grinding, dimpling and Ar$^+$ ion beam thinning until electron transparency. A FEI Tecnai equipped with a field emission gun and a spherical aberration corrector and operating at 200 keV was used for imaging.

RESULTS AND DISCUSSION

The effect of the deposition parameters (such as RF power applied on the HfO_2 cathode, substrate temperature, total plasma pressure, Si surface ratio) on the layers properties was studied according to the range mentioned above.

It was observed that the increase of the total plasma pressure leads, first, to the increase and then to the decrease of deposition rate. The optimal value of P_{total} was found to be 0.04 mbar when deposition rate did not exceed 2 nm/min. Hereafter, the results obtained for the different RFP and T_S will be described for the layers deposited at P_{total}=0.04 mbar.

Usually Hf-O bonds lead to an absorption in the range of 800-600 cm^{-1} as observed by ATR measurement [12,13]. The peak shape and position of the band(s) give the information about the crystalline or amorphous nature of the layers. In the case of the presence of monoclinic HfO_2 phase, the well-defined peaks of LO and TO phonons are detected at 775 cm^{-1} and 685 cm^{-1}. These peaks were predicted theoretically [14] and observed for HfO_2 films [13]. At the same time the observation of only one broad band with the maximum around 700-690 cm^{-1} is usually explained by the amorphous nature of the layers [12]. Unfortunately there are only a few data about vibration bands for HfSiO layers and most of them were obtained for bulk materials [15]. The infrared absorption band corresponding to Si-O-Hf stretching vibration is observed between 888 cm^{-1} and 1004 cm^{-1}. At the same time, for the films prepared by MOCVD, it was also found that the incorporation of silicon into HfO_2 matrix leads to the appearance of the broad band in

70

800-1100 cm^{-1} spectral range, while the peak position shifts from 940 cm^{-1} to 980 cm^{-1} when Si content increases from 10 to 60 at.% [16].

Effect of substrate temperature

Concerning the HfO$_2$ layers, both ATR and XRD study reveal that the increase of T_S at constant RFP leads to the appearance of crystalline phase in the layers (Fig.1). As one can see from Fig.1a, the ATR spectra show the presence of the broad Hf-O vibration band peaked at ~690 cm^{-1} which shape indicates an amorphous structure. Besides, the vibration band in 1260-960 cm^{-1} is also observed. It is usually ascribed to Si-O vibration bands which evidences the formation of a SiO$_2$ interfacial layer during the HfO$_2$ growth. Note that this layer, with different thicknesses, was observed for all pure HfO$_2$ layers whatever RFP and T_S values.

The increase of T_S from 45°C (the minimal temperature of substrate due to plasma effect) to 300 °C does not change the peak position and the shape of Hf-O related band (Fig.1) while Si-O band peak position shifts to higher wavenumber up to 1160 cm^{-1} due to increase of the contribution of LO$_3$ phonon.

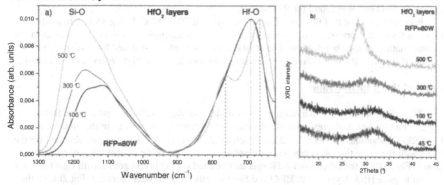

Figure 1. ATR spectra normalized on the intensity of main Hf-O bands (a) and GI-XRD patterns (b) of the as-deposited HfO$_2$ layers. GI-XRD patterns are shifted in vertical direction for their clear presentation. The T_S values are mentioned in the figures. RFP=80 W.

Further increase of T_S, up to 500 °C, leads to the appearance of well-defined absorption peak situated at ~770 cm^{-1} accompanied by the peak at 665 cm^{-1} (Fig.1, curve 3). This indicates the crystallization of HfO$_2$ layer during deposition. Note that the intensity of vibration band in the range 700-690 cm^{-1} is still high probably due to the presence of amorphous HfO$_2$ phase. At the same time a gradual increase of the LO$_3$ phonon contribution for the Si-O band leads to the shift of the Si-O band peak to 1200 cm^{-1}.

Thus, from ATR data, it is possible to conclude that the layers deposited at T_S lower than 300 °C are amorphous, while higher T_S results in the crystallization of the layers and appearance of the monoclinic phase.

XRD study of the layers deposited at different T_S confirms this conclusion. As one can see from corresponding GI-XRD patterns (Fig.1b), the layers grown at T_S=45-300 °C show only one wide peak at 2Θ=31.3-32.0°. The peak position and shape are the characteristics of the amorphous layers. On the other hand the layers could contain also nano-crystalline regions though in such a case the crystallites have not preferable orientation and their sizes do not exceed 2 nm. The increase of T_S up to 500 °C leads to the shift of the peak position to $2\Theta=28.4°$ accompanied by simultaneous increase of its intensity and by the decrease of its full width at half maximum, while the peak at $2\Theta=31.3$-$32.0°$ still gives some contribution to XRD pattern. This could be explained by the simultaneous presence of crystalline and amorphous phases in the layer, that is in an agreement with ATR data obtained for the $T_S=500°C$ (Fig.1a).

Effect of power applied on the cathode

The effect of RFP was also studied for different T_S. The layers deposited at higher RFP and lower T_S are amorphous (Fig.1 and 2). In general, the increase of RFP at a given T_S leads to an increase of the contribution of amorphous phase as well as to the decrease of the crystalline one, that was observed even for the layers deposited at $T_S=500$ °C.

The analysis of ATR spectra versus RFP shows the increase of the intensity of Hf-O vibration band with RFP increase. The highest contribution of Hf-O band to ATR spectra with respect to Si-O one was found for $RFP =60$ W.

It should be noted that the annealing of the layers at 700-900°C during 15 min in a nitrogen flow results in the appearance of a shoulder or well separated peak at ~770 cm^{-1} as well as in the transformation of Si-O band shape. The crystallisation was also confirmed by XRD spectra (not shown) that means that HfO$_2$ layers are not stable, unfortunately, at high temperature annealing.

Effect of the Si surface ratio

The effect of the Si surface ratio, R_{Si}, was studied for $R_{Si}=3$-12%. As one can see from Fig.2a, the increase of R_{Si} results in the decrease of the intensity as well as in the broadening of the XRD peak situated at $2\Theta=31.3$-$32.0°$. Such a behaviour of GI-XRD patterns obtained for the films with the same thickness can be explained by an increase of the contribution of amorphous phase (Fig.2a). On the other hand ATR spectra obtained for the layers grown with $R_{Si}=3$-6% are similar to pure HfO$_2$ layers: both Hf-O and Si-O vibrations band were detected (Fig.2b). At the same time, peak position of Si-O related band differs in comparison with its counterpart observed for the pure HfO$_2$ layer deposited in the same conditions. In this case, the Si-O peak is situated at ~1200 cm^{-1} even for the layer deposited at 100 °C. This can be due to participation of Si ions sputtered from Si chips in the formation of SiO$_x$ interface layer at the initial stage of the growing process. The absence of any well-defined Hf-O related vibration band indicates the amorphous nature of the layers. At the same time the layers deposited with $R_{Si}=12\%$ demonstrate the presence of a broad band peaked at 1050 cm^{-1} and a shoulder at about 900 cm^{-1} which tail expands up to 600 cm^{-1} (Fig.2b). Going further, the annealing of this layer at 800 °C for 15 min (Fig.2b, curve 4) shows only the shift of the broad peak from 1050 cm^{-1} to 1070 cm^{-1} while the other part of ATR spectrum did not change. The shift of the peak should be due to transformation of the Si-O bonds present in the layer and could be explained by the improvement of the interface and film quality [15]. Note that annealing even at 900 °C for 15 min in N$_2$ flow did not lead to the crystallization of the layers since neither a shoulder nor a peak at 775 cm^{-1} were observed after such treatment (not shown). In this case we can conclude that fabrication of the layers from the HfO$_2$ target topped by Si chips results in the growing of thermally stable layers.

At the same time the observation of the broad band in ATR spectrum should be the evidence of the formation of HfSiO matrix.

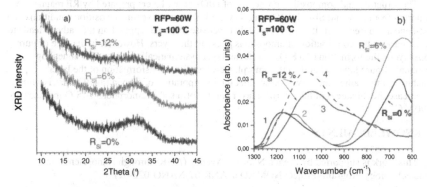

Figure 2. GI-XRD patterns (a) and ATR spectra (b) of as-deposited HfO$_2$–based layers versus R$_{Si}$ mentioned in both figures. T_S=100 °C, RFP=60 W. The dashed curve 4 in b) represents transformation of ATR spectra after annealing of the layer at 800 °C for 15 min in N$_2$ flow.

Figure 3. Cross section TEM images of as-deposited pure HfO$_2$ (left image) and HfSiO (right image) layers. R$_{Si}$=6%, T$_S$=45°C , RFP=60W.

The cross-section TEM images (Fig.3) obtained for as-deposited pure HfO$_2$ and HfSiO layers, at the same RFP and T_S, show that the HfO$_2$ layer is crystalline (left image), while the HfSiO layer is amorphous (right image). Besides, the formation of an interfacial layer was observed. Its thickness was found to be about 4 nm in the case of pure HfO$_2$ layers.The clear contrast of this layer seems to correspond to silicon oxide. At the same time, when HfSiO layer was grown, the decrease of the interfacial layer thickness to 2.5 nm occurred. Besides the lower contrast of the interfacial layer was also found. It is possible to suppose that in this case its composition corresponds to SiO$_x$ or silicate. However, the nature of this layer requires an additional investigation. The electron energy loss spectroscopy (EELS) in the scanning transmission electron microscope (STEM) is in progress to determine precisely its chemical composition.

CONCLUSIONS

The structure and composition properties of HfO_2-based layers prepared by RF magnetron sputtering from pure and Si-rich HfO_2 targets were investigated versus deposition conditions and post-deposition treatment. It was observed that increase of power applied on the cathode leads to the increase of the contribution of amorphous phase in the layers. However, in the case of pure HfO_2 layers, the formation of a SiO_2 interfacial layer systematically occurs. The annealing treatment of pure layers results in the amorphous-crystalline transformation of the layer structure. At the same time is was shown that the incorporation of silicon in HfO_2 matrix allows to decrease the thickness of the interface layer and to form stable amorphous HfSiO matrices up to 900 °C.

ACKNOWLEDGMENTS

This work is supported by French National Agency (ANR) through Nanoscience and Nanotechnology Program (Project NOMAD n°ANR-07-NANO-022-02)

REFERENCES

1. G. D. Wilk, R. M. Wallace and J. M. Anthony, J. Appl. Phys. **89**, 5243 (2001).
2. M. Houssaa, L. Pantisano, L.-A . Ragnarsson, R. Degraeve, T. Schram, G. Pourtois, S. De Gendt, G. Groeseneken and M.M. Heyns, Mat. Sci. Eng. R **51**, 37 (2006).
3. G.Pant, A.Gnade, M.J. Kim, R.M. Wallace, B.E. Gnade; M.A. Quevedo-Lopez, P.D. Kirsch and S. Krishnan, Appl. Phys. Lett **89**, 032904 (2006).
4. J.-H. Kim, V.A. Ignatova and M. Weisheit, Microel. Eng. **86**, 357 (2009).
5. S.K. Dey, A. Das, M. Tsai, D. Gu, M. Flyod, R.W. Carpenter, H. De Waard, C. Werkhoven and S. Marcus, J. Appl. Phys. **95**(9) 5042 (2004).
6. K. Yamamoto, S. Hayashi, M. Kubota and M. Niwa, Appl. Phys. Lett. **81**, 2053 (2002).
7. B.K. Park, J. Park, M. Cho, C.S. Hwang, K. Oh, Y. Han and D.Y. Yang, Appl. Phys. Lett. **80**(13), 2368 (2002).
8. G. He, Q. Fang and L.D. Zhang, Mat.Sci.Semicond.Proc. **9**, 870 (2006)
9. L. Pereira, A. Marques, H. Águas, N. Nedev, S. Georgiev, E. Fortunato and R. Martins, Mat. Sci. Eng. B **109** , 89 (2004).
10. L.-P. Feng, Z.-T. Liu and Y.-M. Sheh, Vacuum **83**, 902 (2009)
11. M.R. Visokay, J.J. Chambers, A.L.P. Rotondaro, A. Shanware and L.Colombo, Appl. Phys. Lett. **80**(17), 3183 (2002).
12. N.V. Nguyen, A.V. Davydov, D.Chandler-Horowitz and M.F. Frank, Appl.Phys.Lett. **87**, 192903 (2005).
13. M.M. Frank, S. Sayan, S. Dörmann, T.J. Emge, L.S. Wielunski, E. Garfunkel, and Y.J. Chabal, Mater. Sci. Eng. B **109**, 6 (2004).
14. X. Zhao, D. Vanderbilt, Phys. Rev . B **65**, 233106 (2002)
15. V. Cosnier, M.Olivier, G. Theret and B. Andre, J.Vac.Sci.Technol.A **19**, 2267 (2001).
16. M.Lui, L.Q. Zhu, G.He, Z.M. wang, J.X. Wu, J.-Y. Zhang, I. Liaw, Q. Fang and I.W.Boyd, Appl. Surf. Sci. **253**, 7869 (2007).

Ferroelectric Memory

Mater. Res. Soc. Symp. Proc. Vol. 1160 © 2009 Materials Research Society   1160-H08-01

Key Technologies for FeRAM Backend Module

Tian-Ling Ren[*], Ming-Ming Zhang, Ze Jia, Lin-Kai Wang, Chao-Gang Wei, Kan-Hao Xue,
Ying-Jie Zhang, Hong Hu, Dan Xie, Li-Tian Liu
Institute of Microelectronics, Tsinghua University, Beijing 100084, China
Tsinghua National Laboratory for Information Science and Technology, Tsinghua University,
Beijing 100084, China
[*]Corresponding author's E-mail: rentl@mail.tsinghua.edu.cn

ABSTRACT

Ferroelectric random access memory (FeRAM) is believed to be the most promising candidate for the next generation non-volatile memory due to its fast access time and low power consumption. Fabrication technologies of FeRAM can be divided into two parts: CMOS technologies for circuits which are standard and can be shared with traditional IC process line, and process relating to ferroelectric which is separated with CMOS process and defined as backend module. This paper described technologies for integrating ferroelectric capacitors into standard CMOS, mainly about modeling of ferroelectric capacitors and backend fabrication technologies. Hysteresis loop of the ferroelectric capacitor is the basis for FeRAM to store data. Models to describe this characteristic are the key for the design of FeRAM. A transient behavioral ferroelectric capacitor model based on C-V relation for circuit simulation is developed. The arc tangent function is used to describe the hysteresis loop. "Negative capacitance" phenomenon at reversing points of applied voltage is analyzed and introduced to the model to describe transient behaviors of the capacitor. Compact equivalent circuits are introduced to integrate this model into HSPICE for circuit simulation. Ferroelectric materials fabrication, electrodes integration and etching are the main technologies of FeRAM fabrication process. An metal organic chemical vapor deposition (MOCVD) process is developed to fabricate high quality $Pb(Zr_{1-x}Ti_x)O_3$ (PZT) films. Pt is known to cause the fatigue problems when used as electrodes with PZT. Ir is used as electrodes to improve the fatigue property of PZT based capacitors, and mechanism of the fatigue is analyzed. Hard mask is used to reduce the size of the capacitors and damage caused in etching process. In our process, Al_2O_3 is developed as hard mask, which simplifies the FeRAM backend integration process.

Key words: ferroelectric, memory, fatigue, modeling, negative capacitance, integration, MOCVD

INTRODUCTION

As Moore's law indicated [1], IC technologies nowadays have scaled down to 32 nanometers [2]. With the shrinking of device feature size, traditional memories, like Flash and DRAM, are facing more and more challenges, such as reliability and power consumption [3]. Memory technologies based on new theories and new materials has attached great attention in recent years, including ferroelectric random access memory (FeRAM) [4], phase change memory (PCM) [5], magnetic random access memory (MRAM) [6], and resistive random access memory (RRAM) [7], and etc. Among the several candidates, FeRAM has attached tremendous interest due to its excellent properties such as non-volatility, low power consumption, high write endurance and high operation speed [8]. Generally speaking, fabrication technologies of FeRAM can be divided into two main parts. The first part is fabrication of COMS circuits, which is standard process and can be shared with normal IC process line. The other part is integration of ferroelectric capacitors with standard CMOS circuits. Usually processes relating to ferroelectric are taken after the fabrication of CMOS transistors, which we define as FeRAM backend module. In this paper, we described some issues about key technologies of realization of FeRAM, including modeling of ferroelectric capacitors and fabrication technologies of FeRAM backend module.

MODELING OF FERROELECTRIC CAPACITORS

As is known, hysteresis is the basic characteristic of ferroelectric material. In FeRAM, two polarization states of ferroelectric material generated in different applied field can be used to distinguish the two data states needed in memory storage. And non-volatility of remnant polarization makes FeRAM non-volatile. In other words, we can say hysteresis property of the ferroelectric capacitor is the basis for FeRAM. A model to describe this characteristic is the key for simulation in the design of FeRAM. Generally speaking, models of ferroelectric capacitors can be divided into two types: physically based models and behavioral models [9]. Behavioral models focus on the electric properties of ferroelectric material from the equivalent circuit point of view, which can be integrated in EDA tools and are easier to realize. In our work, we developed a transient behavioral ferroelectric capacitor model based on C-V relation for FeRAM simulation.

Basic modeling approach of ferroelectric capacitor

Figure 1 illustrates the saturated hysteresis loop of a ferroelectric capacitor under symmetric applied field.

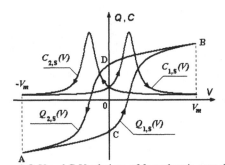

Figure 1 Q-V and C-V relations of ferroelectric capacitors

In our model, the arc tangent function is used to describe hysteresis property:

78

$$Q_{1,S}(V) = \frac{c}{2a}\left[arc\tan\left(\frac{V_m+V_c}{a}\right) - arc\tan\left(\frac{V_m-V_c}{a}\right)\right] + \frac{c}{a}\cdot arc\tan\left[\frac{(V-V_c)}{a}\right] + b\cdot V$$

(1)

$$Q_{2,S}(V) = \frac{c}{2a}\left[arc\tan\left(\frac{V_m-V_c}{a}\right) - arc\tan\left(\frac{V_m+V_c}{a}\right)\right] + \frac{c}{a}\cdot arc\tan\left[\frac{(V+V_c)}{a}\right] + b\cdot V$$

(2)

Where a, b, c are parameters needed to be determined according to the types of ferroelectric materials. C-V relation of ferroelectric capacitors can be obtained by differential Q-V equations:

$$C(V) = \frac{dQ(V)}{dV}$$

(3)

Negative capacitance in ferroelectric capacitor

In the experiment, we find that when the decreasing applied voltage suddenly starts to increase, the polarization keeps decreasing in a short time. During this period, the capacitor seems to have a negative capacitance as shown in Figure 2. The frequency of the triangular signal applied in the test is 100Hz and the amplitude is 5V. We assume that the "Negative capacitance" is caused by the delay of domain reversal after the change of applied voltage. Figure 2(a) is the test result of $SrBi_2Ta_2O_9$ (SBT) capacitor which illustrates negative capacitance. From C-V relation in Figure 2(b), negative peak of capacitance is clearly shown at coercive electric field. Besides, it is found that phenomenon of negative capacitance become more obvious when the frequency of applied voltage increases.

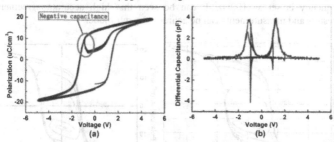

Figure 2 (a) negative capacitance phenomenon in observed in experiment; (b) C-V relation of capacitor obtained by integrating P-V curve in (a).

The "negative capacitance" is considered in our model and is described as follow:

$$C'(V) = C(V)\cdot\left[1 - e^{-k_1\cdot\frac{S}{\ln(S)}\cdot t'}\right] + C(V_x)\cdot k\cdot\ln(S)\cdot\left[-e^{-k_2\cdot\frac{S}{\ln(S)}\cdot t'}\right]$$

(4)

In the equation, k_1, k_2 and k are parameters to be determined, $C(V_x)$ is the capacitance determined from equation (1)~(3) without considering negative capacitance when the applied voltage starts to reverse at V_x, t' is the time from the reversal of applied voltage.

Simulation results

The ferroelectric model considering negative capacitance is implemented in HSPICE in the following schematic diagram:

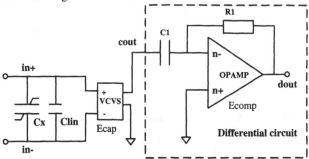

Figure 3 Schematic diagram of model implementation in HSPICE

Ferroelectric capacitor is simulated in the condition that different asymmetric nonperiodic input signals are applied and the result is compared with the test result as shown in Figure 4. Model parameters are extracted by curve fitting the hysteresis loop of 200 nm SBT thin film capacitors that is measured under the voltage at 100 Hz by aixACCT TFAnalyzer 2000. The applied pulse sequence is set as: 0, 5V, -5V, 4V, -4V, 3V, -3V, 2V, -2V, 1V, -1V, 0.5V, -0.5V, and the frequency is set as 100Hz. It can be seen that high accurate agreement between simulation results and measurements can be achieved by our model.

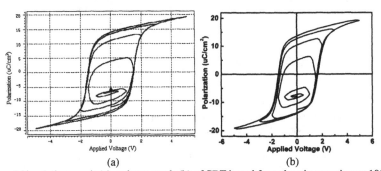

(a)   (b)

Figure 4 Simulation result (a) and test result (b) of SBT based ferroelectric capacitor at 100 Hz

Based on the ferroelectric capacitor model, we simulated the reading access process of a 2T2C FeRAM storage unit. And the result is shown in Figure 5 which proves that our model is capable for the application in FeRAM circuit design.

80

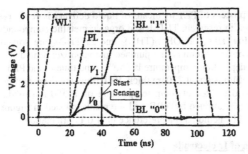

Figure 5 Simulation result of reading cycle including rewriting in a FeRAM 2T2C unit
(Capacitor area is 4 μm^2, CMOS process is of 0.35 μm, bitline parasitical capacitor is considered
as 500 fF, and operation voltage is 5 V)

KEY FABRICATION TECHNOLOGIES OF FERAM BACKEND MODULE

Ferroelectric materials fabrication, electrodes and etching are the main parts to integrate
ferroelectric material with CMOS circuits, and they are key technologies in fabricating FeRAM.

MOCVD process for PZT

Among the several ferroelectric materials applied in FeRAM, such as PZT, SBT, and
$(Bi_{1-x}La_x)Ti_3O_{12}$ [10~12], PZT is found to have outstanding ferroelectric properties to integrate
in FeRAM for its low crystallization temperature and large polarization [13]. Chemical solution
deposition (CSD) and MOCVD are the most used methods in fabricating ferroelectric material.
Since CSD method has critical requirement in planar substrate, it cannot be used in high-density
memory devices [14]. Recent years, MOCVD technology of ferroelectric material, especially
PZT, has been investigated extensively.

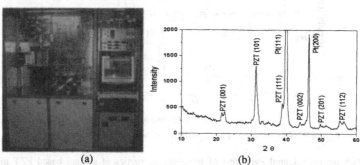

(a)   (b)
Figure 6 (a) MOCVD tool used in experiment; (b) XRD results of PZT films prepared by
MOCVD

81

An MOCVD process of PZT by means of liquid delivery is developed. Figure 6(a) shows the MOCVD tool used in research. THD-based organic metallic compounds are used for metal-organic precursors. $Pb(THD)_2$, $Ti(OPr^j)_2(THD)_2$ and $Zr(THD)_4$ are solved in Tetrahydrofuran (THF). Then the precursor solution is first pumped in a vaporizer heated at 200 °C then delivered into the reactor. The deposition temperature is 650°C. The time of the PZT deposition process is 35 minutes, and PZT film is about 220nm thick. PZT films are deposited on a Pt/Ti (150/20 nm) electrode, which was deposited at room temperature on a 350 nm oxidized Si wafer by DC sputtering. PZT film is examined by XRD and the result is listed in Figure 6(b). It can be seen that PZT prepared in this method has a preference in (1 0 1)-orientation.

Study on integration of Ir electrode

Electrodes are found to have a direct impact on the performance of ferroelectric capacitors [15, 16], including polarization, fatigue and leakage. Among the several electrode materials applied in capacitors of FeRAM, Platinum (Pt) is most widely used for its stability at high temperature and small lattice mismatch with PZT [17]. However, Pt bottom electrode in PZT based capacitors causes severe fatigue problem [18]. Recent years, Iridium is reported to have good barrier effect and has been widely studied for application in FeRAM [19, 20]. In the process of integrating Ir, large stress in Ir thin films which causes the delaminating and cracking of Ir is always a problem [21].

In our experiment, delaminating of as deposited Ir films during PZT fabrication is found. We find that, using the same spin coating method, baking Ir bottom electrodes in the air for 20 minutes at 400 °C before PZT deposition can solve this problem. The baking process is found to cause no obvious change to Ir electrode, and no oxidization happened as shown in Figure 7.

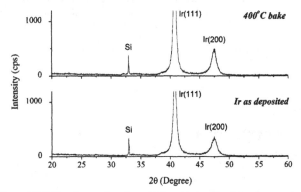

Figure 7 XRD patterns of as deposited and baked Ir electrode samples

Then we investigated the effect of Ir on the properties of sol gel based PZT material compared with Pt bottom electrode. As shown in Figure 8, it is found that PZT prepared on Pt bottom electrode strongly prefers (1 1 1) orientation. However, PZT prepared on Ir prefers (1 0 1) orientation. Besides, in the XRD patterns of PZT prepared on Ir electrode, peaks of IrO_2 is found, which means that Ir electrode is oxidized in the 650 °C PZT crystallization process.

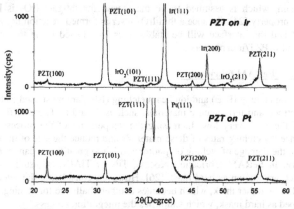

Figure 8 XRD patterns of PZT prepared on Pt and Ir bottom electrode.

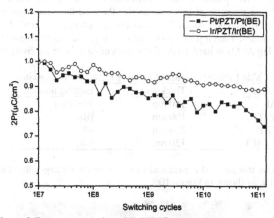

Figure 9 Fatigue properties of Pt/PZT/Pt and Ir/PZT/Ir capacitors.

Pt/PZT/Pt and Ir/PZT/Ir capacitors are prepared. Fatigue properties of the capacitors are studied as shown in Figure 9. An alternating square pulse of 5V at 10MHz is used in the fatigue test. Though Ir electrodes greatly increase the fatigue performance of sol-gel PZT based capacitors, about 10% 2Pr is lost after 1E11 switching cycles. For Pt/PZT/Pt capacitors, it is known that Pb diffusion into the substrate and formation of $PbPt_x$ resulting in Pb vacancies. According to Figure 8, in Ir/PZT/Ir capacitors, IrO_2 layer is found in PZT/Ir interface. We assume that this is caused by the reaction between Ir and oxygen from PZT during the high temperature crystallization process, which may cause oxygen vacancies in PZT. Pb and oxygen vacancies in PZT have been reported to act like holes in p-type semiconductor and electron in n-type semiconductor, respectively [22, 23]. They will trap electrons and holes resulting in the pinning of domain, which is assumed to be the reason for fatigue. IrO_2 is known to have excellent barrier property [24]. So once a thin IrO_2 layer is formed, reaction between Ir and PZT will be relieved and the interface will be stable. This is believed to be the reason for better fatigue property of Ir/PZT/Ir capacitors.

Etching process in FeRAM fabrication

Reactive ion etching (RIE) and ion beam etching (IBE) are most used etching methods in IC process. Electrode and ferroelectric materials, such as Pt and PZT, are difficult to etch. The etching rates of them are very low. Increasing the temperature of the process is an effective method to enhance the etching rates of these materials and reduce the size of the capacitors [25]. In this condition, the damage of etching on photoresist is too great that it cannot be used to form patterns anymore. In FeRAM fabrication process, TiN or TiAlN are usually used instead of photoresist, which are known as hard mask [26]. However, considering the selective rates, TiN and TiAlN need to be rather thick, which becomes another challenge for etching. In our process, Al_2O_3 is developed as hard mask, which simplifies the integration process.

After fabrication of bottom electrode, PZT and top electrode, Al_2O_3 is deposited directly on the top electrode by RF sputtering. An Al_2O_3 target of 99.5% purity was used. The chamber was first evacuated to a pressure of less than 5×10^{-4} Pa. And the deposition was carried out in argon atmosphere at 3 Pa. Then Al_2O_3 is wet etched by phosphoric acid. In this method, the patterns are easily formed. Then top electrode and PZT layers underneath are etched by IBE and RIE respectively using Al_2O_3 as hard mask. The details of each layer are listed in table I.

Table I Process details of forming ferroelectric capacitor

Layer	Thickness	Etching method
Al_2O_3	180 nm	Wet etch
Ir (TE)	100 nm	IBE
PZT	220 nm	RIE
Ir (BE)	150 nm	IBE

It is found that this method is practical to avoid severe damages that the etching process caused on photoresist, as shown in Figure 10.

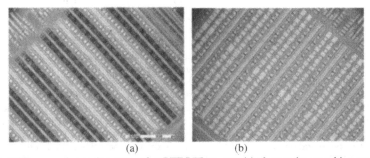

(a)   (b)

Figure 10 Images of capacitor array after PZT RIE process (a) photoresist as etching mask (b) Al_2O_3 as etching mask

Ferroelectric function parts are integrated based on the key technologies of circuit design and fabrication technologies of backend module. Figure 11 illustrated images of a 2Kbit FeRAM developed based on the related key technologies, including the wafer after the fabrication processes, testing die and the packaged chip.

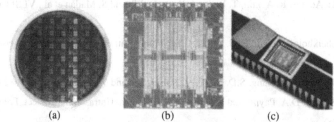

| (a) | (b) | (c) |

Figure 11 Images of a FeRAM (a) wafer; (b) testing die; (c) packaged chip.

CONCLUSIONS

Technology about integrating ferroelectric capacitors with standard CMOS circuits is a key part of FeRAM technologies. Some issues about model and fabrication technologies concerning this part are discussed in this work. A behavior model of ferroelectric capacitors is introduced for FeRAM circuits design. Negative capacitance is illustrated and considered in modeling of ferroelectric capacitors. Then optimized model is implemented in HSPICE. Ferroelectric capacitor and a 2T2C FeRAM unit are simulated using this model. Some fabrication technologies of FeRAM are also presented. MOCVD process of PZT is developed. Ir is investigated and applied in PZT based capacitors as electrodes. The mechanism of fatigue in ferroelectric capacitors is discussed. Etching in the FeRAM backend integration process is studied and Al_2O_3 hard mask is found to be very promising to simplify the process.

ACKNOWLEDGMENT

This work is funded by National Nature Science Foundation of China (90407023).

REFERENCES

[1] G.E. Moore, Proceedings of the IEEE. 86, 82 (1998).

[2] http://www.intel.com/pressroom/archive/releases/20090210corp.htm

[3] R. Bez, E. Camerlenghi, A. Modelli and A. Visconti, Proceedings of the IEEE. 91, 489 (2003).

[4] J.F. Scott and C.A. Araujo, Science. 246, 1400 (1989).

[5] S. Lai, IEDM. 2003, 255.

[6] A. Bette, J. Debrosse, D. Gopl, H. Hoeniqschmid, R. Robertazzi and C. Arndt et al., VLSI Circuits 2003, 217.

[7] A. Chen, S. Haddad, Y.C. Wu, T.N. Fang, Z.D. Lan and S. Acanzino et al., IEDM 2005, 746.

[8] H. McAdams, R. Acklin, T. Blake, J. Fong, D. Liu and S. Madan et al., VLSI Circuits 2003, 175.

[9] A. Sheikholeslami and P. G. Gulak, IEEE Trans. Untrason. Ferroelect. Freq. control. 44, 917 (1997).

[10] B.H. Park, B.S. Kang, S.D. Bu, T.W. Noh, J. Lee and W. Jo, Nature. 401, 682 (1999).

[11] S.K. Dey, D.A. Payne, and K.D. Budd, IEEE Trans. Untrason. Ferroelect. Freq. control. 35, 80 (1988).

[12] K.N. Kim and Y.J. Song, Microelectron. Reliab. 43, 385 (2003).

[13] S. Kobayashi, K. Amanuma and H. Hada, IEEE Electron. Dev. Lett. 19, 417 (1998).

[14] J.K. Lee, M.S. Lee, S. Hong, W. Lee, Y.K. Lee, S. Shin and Y.S. Park, Jpn. J. Appl. Phys. 41 6690 (2002).

[15] M. Angadi, O. Auciello, A.R. Krauss and H.W. Gundel, Appl. Phys. Lett. 77, 2659 (2000).

[16] Y. Chen and P.C. McIntyre, Appl. Phys. Lett. 91, 232906 (2007).

[17] T. Nakamura, Y. Nakao, A. Kamisawa and H. Takasu, Appl. Phys. Lett. 77, 1522 (1994).

[18] K.N. Kim and S.Y. Lee, J. Appl. Phys., 100, 051604 (2006).

[19] K. Aoki, Y. Fukuda, K. Numata and N. Akitoshi, Jpn. J. Appl. Phys. 35, 2210 (1996).

[20] K. N. Kim and Y.J. Song, Microelectron. Reliab. 43, 385 (2003).

[21] Y.J. Song, H.H. Kim, S.Y. Lee, D.J. Jung, B.J.Koo and J.K. Lee, Appl. Phys. Lett. 76, 451 (2000).

[22] R. Gerson and H. Jaffe, J. Phys. Chem. Solids. 24, 979 (1963).

[23] S. Madhukar, S. Aggarwal, A.M. Dhote, R. Ramesh, A. Krishnan and D. Keeble, J. Appl. Phys. 81, 3543 (1997).

[24] C.U. Pinnow, I. Kasko, C. Dehm, B. Jobst, M. Seibt and U. Geyer, J. Vac. Sci. Tech. B 19, 1857 (2001).

[25] S. Marks, J.P. Almerico, M.K. Gay and F.G. Celii, Integr. Ferroelect. 59, 333 (2003).

[26] S.Y. Lee and K. Kim, IEDM 2002, 547.

Mater. Res. Soc. Symp. Proc. Vol. 1160 © 2009 Materials Research Society   1160-H08-03

Mechanical Constraint and Loading on Ferroelectric Memory Capacitors

I. Pane[1], N.A. Fleck[2], D.P. Chu[2] and J.E. Huber[3]
[1]Department of Civil Engineering, Bandung Institute of Technology
Ganesha 10, Bandung 40132, Indonesia
[2]Department of Engineering, University of Cambridge
Trumpington Street, CB2 1PZ, UK
[3]Department of Engineering Science, University of Oxford
Parks Road, OX1 3PJ, UK

ABSTRACT

The influence of mechanical constraint imposed by device geometry upon the switching response of a ferroelectric thin film memory capacitor is investigated. The memory capacitor was represented by two-dimensional ferroelectric islands of different aspect ratio, mechanically constrained by surrounding materials. Its ferroelectric non-linear behaviour was modeled by a crystal plasticity constitutive law and calculated using the finite element method. The switching response of the device, in terms of remnant charge storage, was determined as a function of geometry and constraint. The switching response under applied in-plane tensile stress and hydrostatic pressure was also studied experimentally. Our results showed that (1) the capacitor's aspect ratio could significantly affect the clamping behaviour and thus the remnant polarization, (2) it was possible to maximise the switching charge through the optimisation of the device geometry, and (3) it is possible to find a critical switching stress at zero electric field and a critical coercive field at zero residual stress.

INTRODUCTION

A thin film ferroelectric memory capacitor normally consists of a ferroelectric film of thickness about 100 nm, sandwiched between electrodes, each of thickness about 100 nm. The device is embedded within a silicon oxide passivation layer, grown on a silicon substrate. The cross-section of a typical ferroelectric capacitor is shown in Fig. 1. While in use, an electric field is applied and the ferroelectric layer undergoes polarisation switching. This allows for storing binary data. Individual bits of binary data are represented by the local polarisation of the film, with upward and downward polarisation corresponding to the two binary states.

Ferroelectric Random Access Memory (FeRAM) offers features such as fast reading and writing operation, is highly re-writable, scalable, and compatible with Si technology - either as a stand-alone chip or embedded within an on-chip system. It offers a huge potential for the market of low-power non-volatile memory devices. Continuous reduction of the CMOS feature size [1] demands a corresponding reduction in the size of the ferroelectric capacitor. Consequently the capacitor must be designed for the maximum possible polarisation switch, to provide a sufficient charge flow for detection of the memory state. Nonlinear ferroelectric polarisation switching is the main operating mechanism of the ferroelectric memory capacitor. Yet, the nonlinear ferrolectric behavior especially under the influence of mechanical constraint has not been throughly investigated. The aim of the current study is to investigate such influence upon the nonlinear switching response of a ferroelectric capacitor.

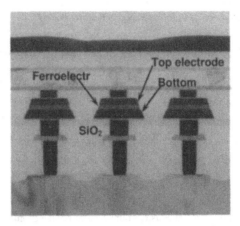

Figure 1. Cross-sectional TEM image of integrated ferroelectric thin film capacitors (with permission from Seiko-Epson, Japan).

MODELING AND SIMULATED RESPONSE

Model

The ferroelectric material is modeled as a single crystal by the rate dependent constitutive law of Huber and Fleck [2]. The model represents the total strain ε_{ij} and electric displacement D_i as the sum of remnant parts (ε_{ij}^r, P_i) and reversible parts ($\varepsilon_{kl} - \varepsilon_{kl}^r$, $D_i - P_i$). The reversible strain and electric displacement are related to the stress σ_{ij} and electric field E_i by linear piezoelectric relations, so that

$$\sigma_{ij} = c_{ijkl}\left(\varepsilon_{kl} - \varepsilon_{kl}^r\right) - e_{kij}E_k \tag{1}$$

$$D_i = e_{ikl}\left(\varepsilon_{kl} - \varepsilon_{kl}^r\right) + \kappa_{ik}^\varepsilon E_k + P_i \tag{2}$$

where $e_{kij} = c_{ijmn}d_{kmn}$, $\kappa_{ik}^\varepsilon = \kappa_{ik}^\sigma - d_{irs}c_{pqrs}d_{kpq}$, κ_{ik}^σ is the dielectric permittivity tensor, c_{ijkl} is the elastic stiffness tensor, and d_{kij} is the piezoelectric tensor. A representative volume of a single tetragonal ferroelectric crystal has $M = 6$ crystal variants or domain types corresponding to the six polarization directions shown in Fig. 2. Let the Ith domain have volume fraction c_I and polarization direction n_i. For simplicity, the elastic stiffness c_{ijkl} and dielectric permittivity κ_{ik}^σ are taken to be isotropic and do not vary from domain to domain. Then c_{ijkl} depends only upon the shear modulus μ and Poisson ratio ν, while the isotropic permittivity tensor scales with a single parameter κ. The piezoelectric tensor of the Ith domain is given by

$$d_{ijk}^I = d_{33}n_i n_j n_k + d_{31}\left(n_i\delta_{jk} - n_i n_j n_k\right) + d_{15}\left(\delta_{ij}n_k - 2n_i n_j n_k + \delta_{ik}n_j\right) \tag{3}$$

where d_{33}, d_{31}, and d_{15} are material coefficients. The remnant strain and polarisation of the Ith domain are $\varepsilon_{ij}^{r,I} = \varepsilon_0 \left(n_i n_j - \delta_{ij} \right)/2$ and $P_i^I = P_0 n_i$ where ε_0 and P_0 are material constants defining the magnitude of remnant strain and polarisation in the domain. The model assumes that the stress and electric field are uniform over the representative volume, while the remnant strain and polarization are given by volume averages. Consequently, the macroscopic piezoelectric tensor, d_{ijk}, equals to $\sum_{I=1}^{M} c_I d_{ijk}^I$. Switching systems (also known as transformation systems) allow for the transformation of one variant type (J) into another type (I). The total number of switching systems that can be active simultaneously in a tetragonal crystal is 15.

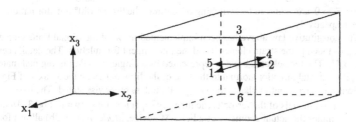

Figure 2. Polarisation directions in a [100]-oriented tetragonal crystal.

Full switching of the αth transformation generates a change of remnant strain by $\Delta\varepsilon_{ij}^{r,\alpha} = \varepsilon_{ij}^{r,I} - \varepsilon_{ij}^{r,J}$, of remnant polarisation by $\Delta P_i^{\alpha} = P_i^I - P_i^J$, and of the piezoelectric tensor by $\Delta d_{ijk}^{\alpha} = d_{ijk}^I - d_{ijk}^J$. Using these quantities, the thermodynamic driving force G^{α} for the αth transformation, converting variant J to variant I is:

$$G^{\alpha} = \sigma_{ij}\Delta\varepsilon_{ij}^{r,\alpha} + E_i\Delta P_i^{\alpha} + \sigma_{ij}\Delta d_{ijk}^{\alpha}E_k \tag{4}$$

Letting the rate of change of domain volume fraction c_I due to the αth transformation be $\dot{f}^{\alpha}$, the rate dependent switching process is simulated using the following equation,

$$\dot{f}^{\alpha} = \dot{f}_0 \left|\frac{G^{\alpha}}{G_c^{\alpha}}\right|^{m-1} \frac{G^{\alpha}}{G_c^{\alpha}} \left(\frac{c_I}{c_0}\right)^{1/k} \tag{5}$$

where $\dot{f}_0$ is a reference switching rate, G_c^{α} is the critical value of G^{α} at which there is a rapid increase in transformation rate, and c_0 is the initial volume fraction of crystal variant I. The rate exponents m and k control the rapidity of the onset of switching when $G^{\alpha} \cdot G_c^{\alpha}$ and the saturation of switching when $c_I \cdot 0$ respectively. The resulting rates of change of remnant strain, remnant polarisation, and of the piezoelectric tensor are given by

$$\sum_{\alpha=1}^{N} \dot{f}^{\alpha}\Delta\varepsilon_{ij}^{r,\alpha} \,, \qquad \dot{P}_i = \sum_{\alpha=1}^{N} \dot{f}_0^{\alpha}\Delta P_i^{\alpha}, \qquad \dot{d}_{ijk} = \sum_{\alpha=1}^{N} \dot{f}^{\alpha}\Delta d_{ijk}^{\alpha} \tag{6}$$

In the rate independent limit (m · 1), G_c^α is the energy barrier for αth transformation. The value of G_c^α for 90° and 180° switching systems can be expressed in terms of the electric field strength required to cause switching in a single crystal. Assume that the crystal has the six possible polarisation directions shown in Fig. 2, and electric field is applied parallel to the x_3 direction. In the rate independent limit, $G_c^{180} = 2E_{180}P_0$ for 180° switching and $G_c^{90} = \sqrt{2}E_{90}P_0$ for 90° switching, where E_{180} and E_{90} are the electric field required for 180° and 90° switching, respectively. The relative values of the barriers to 180° and 90° switching may in practice be varied by doping or by control of the material processing route. We introduce the parameter $\bar{r} = G_c^{180}/(G_c^{180} + G_c^{90})$ where $0 \le \bar{r} \le 1$ to represent a range of material behavior. When $\bar{r} = 0$, 90° switching dominates, while $\bar{r} = 1$ corresponds to dominance of 180° switching. Throughout, $\bar{r}$ is varied by changing the value of G_c^{90} while holding $G_c^{180} = 2E_{180}P_0$ constant. The limits $\bar{r} \to 1$ and $\bar{r} \to 0$ are achieved by switching off entirely the 90° or 180° transformation processes respectively.

The constitutive law above has been implemented in two dimensional finite elements (FE) using 6-noded plane strain elements and the rate tangent formulation. The details can be found in [3,4]. The ferroelectric solid is represented by a single crystal of tetragonal material with the [100] crystal direction normal to the face of the film and along the x_3-axis of Fig. 1. Calculations are carried out in plane strain ($\varepsilon_{22} = 0$) unless otherwise stated. The two dimensional (2D) models of the ferroelectric memory capacitor are shown in Fig. 3. *Model A* (Fig. 3a) resembles the actual geometry of a FeRAM device. *Model B* (Fig. 3b) allows for an examination of the significance of the mechanical constraint against switching imposed by the electrodes. *Model C* (Fig. 3c) is typical of a laboratory test structure and allows for an assessment of the significance of the SiO₂ passivation upon switching, when compared with geometry A. *Model D* (Fig. 3d) is an idealised laboratory test structure with electrode layers much thinner than the ferroelectric layer.

The thickness of the ferroelectric layer is h. The Si substrate is of thickness $H_{Si} = 15h$ and the passivation layer, where present, has total thickness $H = 20h$ with the capacitor embedded a distance $3h$ from the substrate, as shown in Fig. 3. These choices represent thick SiO₂ passivation on a thick Si substrate as is common in real devices. The ferroelectric capacitor is $2w$ wide and the total width of the FE model is $2W$ with $W = 10w$ such that the capacitor is remote from the edges of the model. Perfect mechanical bonding at all interfaces is assumed and all layers have isotropic elastic properties, with Young's modulus E and Poisson ratio ν as listed in Table 1. The electrodes, where present, have thickness h and are assumed to be perfectly conducting, resulting in a constant voltage boundary condition on the ferroelectric-electrode interface. For simplicity, the SiO₂ layer and Si substrate are taken to be perfectly insulating, weak dielectrics, so that charge free boundary conditions exist at the ferroelectric-SiO₂ interface, with zero electric displacement outside the ferroelectric layer. Parameters for the ferroelectric single crystal are chosen to be representative of a soft PZT composition (see Table 2). In this work, we simplify the initial state by assuming equal volume fractions $c_I = c_0 = 1/6$ of each domain type, and zero residual stress. In simulations, a uniform, time varying, electrical potential difference $\Phi(t)$ between the top and bottom faces of the ferroelectric capacitor are applied. The magnitude of $\Phi(t)$ is made large enough to produce an average electric field of more than twice the coercive field.

Simulated Response

Before exploring the role of device geometry it is instructive to refer to the following idealised cases studied earlier [4]:

- *3D constraint,* representing a fully encapsulated and rigidly constrained 3D device.
- *2D constraint,* representing an unpassivated film bonded to a substrate.
- *1D constraint.* This mimics a free-standing film, but constrained along a single edge to behave in plane-strain.
- *0D constraint.* A material element without mechanical constraint, representing a free-standing film.

These are limiting cases which can all be represented in a one dimensional model and do not require definitions of electrode layers, passivation layers or susbtrate. Different constraints are simply realised by varying the mechanical boundary conditions.

Table 1. Elastic properties of the layers.

Material	Young's modulus E (GPa)	Poisson ratio ν
PZT ferroelectric	160	0.3
Platinum (Pt) electrodes	160	0.38
SiO$_2$ passivation	50	0.2
Si substrate	160	0.2

Table 2. Material parameters used in the simulations.

Parameter	Value	Unit
Remanent Polarization (P_0)	0.5	Cm^{-2}
Remnant Strain (ε_0)	1%	-
d_{33}	300×10^{-12}	mV^{-1}
d_{31}	-135×10^{-12}	mV^{-1}
d_{15}	525×10^{-12}	mV^{-1}
κ	5.0×10^{-9}	Fm^{-1}
Creep exponent m	5.0	-
Saturation exponent k	1.0	-
Reference rate $\dot{f}_0$	2.0	s^{-1}
Switching field E_{180}	2.0	MVm^{-1}

To characterise the response, a geometry parameter $\bar{w} = w/(w+h)$ is introduced, such that $0 \le \bar{w} \le 1$. The limit $\bar{w} = 0$ represents a device with height much greater than its width. By contrast, a device with $\bar{w} = 1$ has width much greater than its thickness. Saturated polarisation hysteresis loops for models/geometries A-D and for $\bar{r} = 0.5$ and $\bar{w} = 0.5$ are given in Fig. 4. The responses of material elements under 1D and 3D constraint are included. Since the electric field and electric displacement are non-uniform in these geometries, the normalised mean electric displacement q_s / P_0 is shown versus normalised mean electric field Φ / hE_{180}, where q_s is the mean charge density on the top electrode-ferroelectric interface. Recall that there is a diminishing effect of mechanical constraint upon the switching response as $\bar{r}$ is increased since 90° switching is accompanied by a change in strain, whereas 180° switching does not. Consequently, when $\bar{r} = 1$, the four finite geometries have an almost identical response to that of a material element under 3D constraint. For $\bar{r} < 1$, the remnant charge density q_s decreases as the amount of constraint increases through the sequence of geometry D through to A; in each of these cases the response of the finite island is intermediate between that for a material element under 1D constraint and under 3D constraint.

The effect of SiO$_2$ passivation can be seen by comparing the hysteresis loops of models A and C (electrode layers present), or of models B and D (electrode layer absent). Similarly, the effect of electrode layers can be seen by comparing the hysteresis loops of models A and B for a capacitor with SiO$_2$ passivation, or by comparing the hysteresis loops of models C and D for a capacitor without passivation. For $\bar{r} = 0.5$, it is clear that the constraint provided by the passivation layer is enhanced by the presence of the electrodes.

The plot of remnant charge density q_s as a function $\bar{w}$ for $\bar{r} = 0.5$ is shown in Fig. 5. Remnant charge densities for the four limit cases of constraint are also shown in the figure. Generally, for $\bar{r} < 1$, there is a monotonic drop in q_s with increasing $\bar{w}$ for geometries C and D (passivation layer absent), whereas q_s first increases and then decreases with increasing $\bar{w}$ for geometries A and B (passivation layer present). This can be explained as follows. The geometries A and B are constrained along the x_3 direction by the encapsulation whereas geometries C and D are unconstrained in this direction. Therefore, at both limits $\bar{w} \to 0$ and $\bar{w} \to 1$ the geometries A and B behave like the fully constrained material element. At intermediate values of $\bar{w}$, the level of constraint upon geometries A and B is reduced and consequently q_s is greater than that of the material element under 3D constraint. We note from Fig. 5 that q_s has a peak value at $\bar{w} \approx 0.4$. In contrast, the geometries C and D behave like a material element under 1D constraint as $\bar{w} \to 0$, and under 2D constraint as $\bar{w} \to 1$. The mechanical constraint upon geometries C and D is always less than that upon A and B, and consequently their q_s are greater for any given value of $\bar{w}$.

It is instructive to plot (see Fig. 6) the dependence of q_s upon $\bar{r}$ for the finite island geometries with $\bar{w} = 0.5$ and to compare with the response of constrained material elements. A monotonically decreasing response for q_s with increasing $\bar{r}$ is evident in all cases. Consistent with the results already shown in Fig. 5, the presence of 90° switching confers sensitivity of q_s to the degree of constraint. 180° switching dominates for $\bar{r}$ above about 0.7 and the responses for the various geometries then converge to a common value.

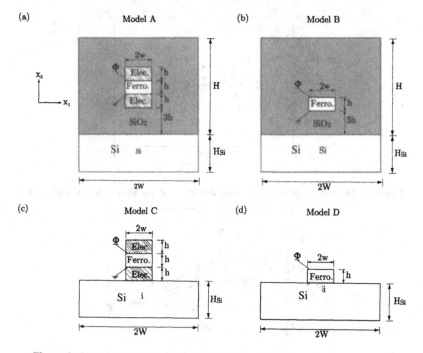

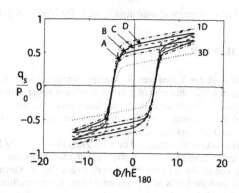

Figure 3. Idealised two-dimensional models (A to D) of ferroelectric capacitors.

Figure 4. Hysteresis loops of Models A to D, for $\bar{w}$ =0.5 and $\bar{r}$ =0.5.

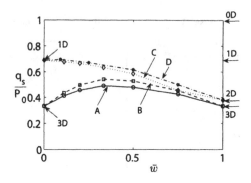

Figure 5. Dependence of remnant charge density on $\bar{w}$ for Models A to D with $\bar{r} =0.5$.

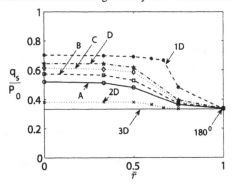

Figure 6. Dependence of remnant charge density on $\bar{r}$ for Models A to D with $\bar{w} =0.5$.

EXPERIMENTAL BEHAVIOR

The experimental work conducted provides the measured results for the polarization and coercive voltage changes in a PZT thin film of 130 nm thickness under uniform in-plane tensile stress and hydrostatic pressure. While the effect of applied stresses are not exactly the same as the influence of mechanical constraint, they allow for a similar mechanical effect on the switching response. The detailed work can be found in [5,6].

Briefly, the experiments were conducted on a $Pb(Zr_{0.45}Ti_{0.55})O_3$ or PZT thin film for the measurement under the uniform in-plane tensile stress [5] and for the measurement under hydrostatic pressure [6]. The structure of the PZT thin film capacitor is shown in Fig 7. In both cases, the annealed films with thickness of 130 nm were in the polycrystalline tetragonal state with a dominant orientation of (111). A uniform tensile stress was applied using a four-point bending test rig. Hydrostatic pressure was applied by means of a high pressure cell. While the crystal orientation of the ferroelectric in the simulation differs from the ones measured

94

experimentally, similar features corresponding to mechanical effects upon the switching response are expected.

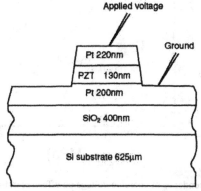

Figure 7. Structure of a PZT thin film capacitor.

In order to compare the results between the effect of hydrostatic pressure σ_{hy} and that of the in-plane stress σ , both stress states are converted to an equivalent out-of-plane stress $\sigma_{\perp}$ such that $\sigma_{\perp} = (1 - 2v)\sigma_{hy}$ and $\sigma_{\perp} = v\sigma$ with a value of Poisson ratio $v = 0.3$. A detailed explanation of how this result is obtained can be found in [6]. The polarization changes in terms of the equivalent out-of-plane stress are shown in Fig. 8 for both hydrostatic pressure and in-plane loading cases. We can see that the two results agree very well with each other, which shows that the nature of the polarization change in these two different stress conditions is essentially the same. Fig 8. also allows for the determination of critical stresses of $\sigma_{\perp c} = 2.9$ GPa and $\sigma_{\perp c} = 1.4$ GPa, respectively.

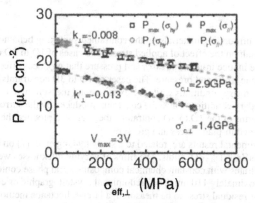

Figure 8. Dependence of maximum polarization P_{max} and remnant polarization P_r on effective out-of-plane stress.

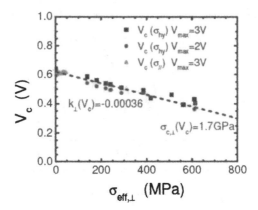

Figure 9. Dependence of coercive voltage on effective out-of-plane stress.

The value of P_{max} and $\sigma_{\perp c}$ are dependent on the maximum applied voltage V_{max}. The experiment was repeated with different maximum applied voltages equal to and greater than 2 V. By extrapolation to $V_{max}=0$ V a critical stress $\sigma_{\perp c}=1.6$ GPa, which is slightly higher than the critical stresses for remnant polarization P_r of 1.4 to 1.5 GPa is obtained. It is reasonable to believe that as the applied voltage is reduced to zero the critical stress becomes the intrinsic critical stress of material.

In addition, it is found from Fig. 8b that the coercive voltage V_c reduces linearly as the out-of-plane stress increases. The critical stress for V_c is deduced to be 1.7 GPa with a material intrinsic coercive voltage of 0.62 V, as shown in Fig. 9. Our results indicate that the maximum applied voltage has little effect on the coercive field.

DISCUSSION

The role of surrounding materials on the contrained switching behavior can be related to the experimental result on the effect of applied stress. The cases of 1D to 3D constraints defined and simulated above induce in-plane stresses and pressure that produces effects similar to those of the applied in-plane stress and pressure. The cases of 1D to 3D constraints however are based on displacement boundary conditions while the experimental measurement uses the stress/traction boundary condition. Since the constraint produced by the surrounding layers have been explained in terms of the 1D to 3D contraints, they can also produce the same effect as that produced by an applied in-plane stress and pressure.

These experimental results are related to a study by Ong et al. [7] on the effect of residual stress in $Pb(Zr_{0.53}Ti_{0.47})O_3$ films. The films of varying thicknesses were deposited onto Pt/Ti/SiO2//Si substrates with constant chemical composition and phase content. The average grain size was approximately 110 nm. The predominant crystallographic orientation was (111). In order to allow for residual stress to be measured, a laser-reflectance method was used to measure wafer curvature before and after PZT deposition and heat treatment. Among the features observed in the experiment is that reactions that occurred on heating Pt/Ti/SiO2//Si affected the baseline curvature, and these effects had to be accounted for in the calculation of residual stress. Tensile stresses between 180 and 150 MPa were calculated from the Stoney equation for films with thicknesses ranging between 95 and 500 nm. It can be expected that as the thickness varies the remnant polarisation and the coercive field also change. This implies that the thickness of the ferroelectric film may be controlled to tailor the switching response.

CONCLUSIONS

From the simulated switching behavior of finite ferroelectric islands with mechanical constraint we gain some useful qualitative insight into the performance of ferroelectric memory devices. Constraint upon 90° switching arises from a number of device features: the electrodes, encapsulation and the substrate. The impact of mechanical constraint is strongly influenced by the relative ease of 90° and 180° switching. As 90° switching is reduced the mechanical constraint becomes less significant. In the limit of purely 180° switching, the response is almost independent of constraint and device geometry. From the standpoint of device design, it is shown that an embedded ferroelectric capacitor may exhibit markedly different response from that of the extended regions of thin film that are commonly used as laboratory test specimens. The study also indicates an optimal geometry of ferroelectric device that maximises the achievable remnant polarisation.

Polarization change in a PZT thin film of 130 nm in thickness under in-plane tensile stress and hydrostatic pressure has been measured. The polarization decreases with the increase in applied pressure, similar to the case of in-plane tensile stress. Using the equivalent out-of-plane stress, the polarization changes in the cases of in-plane tensile stress and hydrostatic pressure were found to be of the same nature. The experimental study obtains a material intrinsic critical stress of 1.6 GPa under uniaxial loading and intrinsic coercive voltage of 0.62 V at the stress-free state. The effect of applied stress upon switching has also been related to the film thickness. As the film thickness changes, the remnant polarisation and the coercive filed also change. The design implication is that the thickness of the ferroelectric film may be controlled to produce the desired switching response. One design approach is to relate the remnant polarisation and coercive voltage to the residual stress and to relate the residual stress to the film thickness.

REFERENCES

1. ITRS, International Technology Roadmap for Semiconductors (2007).
2. J.E. Huber and N.A. Fleck, *J. Mech. Phys. Solids* 49, 785 (2001).
3. I. Pane, N.A. Fleck, J.E. Huber, D.P. Chu, *Int. J. Solids Struct.* 45, 2024 (2008).
4. I. Pane , N.A. Fleck, D.P. Chu, J.E. Huber, *Eur. J. Mechanics A/Solids* 28, 195 (2009).
5. H. Zhu, D. P. Chu, N. A. Fleck, I. Pane, J. E. Huber, and E. Natori, *Integr. Ferroelectrics* 95, 117 (2007).
6. H. Zhu, D. P. Chu, N.A. Fleck, S. E. Rowley, and S.S. Saxena, *J. Applied Physics* 105, 061609 (2009).
7. R.J. Ong, D.A. Payne, and N.R. Sottos, *J. Amer. Ceramics Soc.* 88, 2839(2005).

Resistive Switching RAM I

Mater. Res. Soc. Symp. Proc. Vol. 1160 © 2009 Materials Research Society   1160-H09-03

A Simulation Model of Resistive Switching in Electrochemical Metallization Memory Cells

S. Menzel[1,3], B. Klopstra[1,3], C. Kügeler[2,3], U. Böttger[1,3], G. Staikov[2,3] and R. Waser[1,2,3]
[1] Institut für Werkstoffe der Elektrotechnik II, RWTH Aachen, Aachen, 52074, Germany
[2] Institut für Festkörperforschung, Forschungszentrum Jülich GmbH, Jülich, 52425, Germany
[3] JARA - Fundamentals of Future Information Technology

ABSTRACT

In the present study a simulation model for set operation in electrochemical metallization memory cells was developed to obtain a better understanding of the physical processes involved in resistive switching. The set operation based on filamentary growth within a solid electrolyte was simulated using continuity equation to address the electric properties and a level set method to track the boundary of the filament. FEM simulations were performed using Comsol Multiphysics. The results showed good agreement to experimentally observed I/V - curves for set operation. Furthermore, it could be demonstrated that only one filament is responsible for set operation. Based on this FEM model a simplified resistor based 1D model was developed, showing good agreement to each other. As a refinement, Butler-Vollmer boundary conditions were introduced. This nonlinearity led to an exponential dependency between switching time and switching voltage, which is also observed in experiment.

INTRODUCTION

The Electrochemical Metallization Cell (ECM) is a promising candidate for future non-volatile random access memory overcoming the limits of DRAM (volatile) and Flash (slow). The storage principle is based on change of cell resistance induced by electro-chemical driven growth and rupture of a copper or silver filament in an insulating matrix. This kind of switching was found in several materials such as AgGeSe, CuGeS, SiO_2, WO_3 and MSQ [1-4].

In the present study modeling is based on a copper (Cu) filament without loss of generality. During write (set) operation Cu is oxidized at the corresponding electrode and Cu ions are driven out of the Cu anode into the insulating matrix due to the applied field, whereas the insulating matrix serves as solid electrolyte. The Cu ions migrate towards the cathode. At the cathode electrochemical reduction occurs, and deposition of metallic Cu takes place. Fast drift paths in the solid electrolyte matrix or preferred nucleation sites (surface inhomogenities) at the boundary lead to filamentary growth. This growing Cu filament finally reaches the anode and switches the device to a low resistance state. In the present study modeling is based on a Cu filament without loss of generality.

THEORY

FEM Simulation Model

A simulation model for set operation needs to account for the electrical properties, the Cu ion migration as well as filamentary growth. To simplify the model the dissolution of the anode is neglected, since the active volume of the anode is large compared to the volume of the filament. In addition Cu ion concentration gradients within the solid electrolyte matrix are neglected, resulting in zero space charge. It is further assumed that the conductivity of the solid electrolyte σ_{se} is ionic in nature and electronic contributions are negligible (see eq. (1)). Hence, it is sufficient to solve the well-known Poisson equation (eq. (2)) to account for the electric properties as well as Cu ion migration.

$$\sigma_{se} = 2c_{Cu^{2+}} e \mu_{Cu^{2+}} \tag{1}$$

$$\nabla \sigma \nabla V = 0 \tag{2}$$

Here c_{Cu2+} is the concentration of copper ions, μ_{Cu2+} the mobility of copper ions in the solid electrolyte and e is the elementary charge. The charge number of copper ions was set to two.

The level set method [5] is used to track the boundary of the growing filament. Mathematically, the level set function ψ describes a function that rises smoothly from 0 to 1 within a certain distance - equivalent to the interface thickness γ. In the present model a value of the level set function of 0 corresponds to the solid electrolyte, whereas a value of 1 represents the filament. A value of 0.5 represents the boundary of the growing filament. The movement of the boundary is described by equation (3),

$$\frac{\partial \psi}{\partial t} + \vec{u} \nabla \psi = \gamma \nabla \left[\zeta \nabla \psi - \psi (1 - \psi) \frac{\nabla \psi}{|\nabla \psi|} \right] \tag{3}$$

whereas the right hand side of the equation controls the stability of the interface with ζ being a re-initialization parameter. The velocity u describes the movement of the interface in all directions. Assuming that all copper ions react to metallic copper at the interface, the growth velocity u can be calculated from the ionic current density j_{Cu2+}, the atomic mass of copper M_{Cu} and the density of copper ρ_{Cu} to:

$$\vec{u} = \vec{j}_{Cu^{2+}} \frac{M_{Cu}}{2e\rho_{Cu}} \tag{4}$$

By using the level set equation it becomes necessary to introduce a seed layer at the cathode which can grow during simulation. Preferred nucleation sites at the boundary are modeled using surface protuberances in the seed layer. The set of equations (1)-(4) is completed by appropriate boundary conditions, enabling current driven or voltage driven simulations with current compliance.

The geometry of the ECM cells consists of a 20 nm thick inert bottom electrode (e.g. Pt), a 2 nm thick copper seed layer, a 58 to 80 nm thick solid electrolyte layer and a 20 nm thick copper top electrode. The diameter of the cell is 80 nm in 2D axial symmetric simulations and a 20 nm depth is assumed in 2D simulations.

1D resistor based model

Based on the FEM model a simplified 1D model was developed. The shape of the filament was approximated as a cylinder with a lateral area A_{fil}. Therefore the 2D axial symmetric model can be reduced to an equivalent electrical circuit shown in Figure 1. As the

102

solid electrolyte has a conductivity some orders lower than that of the filament, the model can be further simplified to four resistors in series shown in Figure 2.

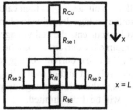

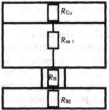

Figure 1: Resistor based 1D model   **Figure 2**: Simplified resistor based 1D model

Using the simplified resistor model in Figure 2, the cell resistance can be calculated by equation (5) with A_{se} being the effective area of the solid electrolyte above the filament.

$$R_{Cell}(x) = R_{TE} + R_{BE} + R_{se1}(x) + R_{fil}(x) = R_{TE} + R_{BE} + \rho_{se}\frac{x}{A_{se}} + \rho_{fil}\frac{(L-x)}{A_{fil}} \quad (5)$$

The resistances of the filament R_{fil} and the solid electrolyte region R_{se1} are both dependent on x with x describing the position of the interface between solid electrolyte and filament. The position of the interface x can be calculated as in the FEM model by the growth velocity (eq. 4). If only ionic currents are present in the solid electrolyte and deposition only takes place at the top of the filament, this leads to the ordinary differential equation (7). The differential equation can be solved analytically or numerically. With known position x the cell resistance as well as the cell voltage or cell current can be easily calculated.

$$u = \frac{\partial x}{\partial t} = -\frac{M_{Cu}}{2e\rho_{Cu}} j_{Cu^{2+}} = -\frac{M_{Cu}}{2e\rho_{Cu}A_{fil}} I_{ges}(t) = -\frac{M_{Cu}}{2e\rho_{Cu}A_{fil}} \frac{V_{cell}(t)}{R_{cell}(x)} \quad (7)$$

Butler-Vollmer boundary conditions

The current density across the electrode interface involving the redox reactions was modeled by the Butler-Vollmer (BV) equation (equation (8)). It describes the current as a function of the overvoltage η an asymmetry factor a, the exchange current density j_0, the number of electrons n involved in the electrode reaction and thermal voltage V_T.

$$J = j_0\left[\exp\left(na\frac{\eta}{V_T}\right) - \exp\left(-(1-a)n\frac{\eta}{V_T}\right)\right] \quad (8)$$

For simplicity the asymmetry factor a is set to 0.5. In this case the BV equation can be simplified to equation (9) allowing for explicit calculation of the overvoltage (eq. 10).

$$J = 2j_0\sinh\left(\frac{n\eta}{2V_T}\right) \quad (9)$$

103

$$\eta = \frac{2V_T}{n} \sinh^{-1}\left(\frac{J}{2j_0}\right) \tag{10}$$

At both electrodes an overvoltage occurs. The overall cell voltage V_{Cell} is then calculated by adding the absolute two overvoltages $\eta_{cathode}$ and η_{andode} to the voltage drop due to ohmic losses of the electrodes and solid electrolyte, respectively. At low currents the cell voltage is determined by the two overvoltages at the interfaces. In this regime the current across the interface is electron - transfer limited. According to equation (9) a linear current voltage relation is expected for low overvoltages and very low currents, respectively. At higher overvoltages this relation becomes exponential. If the current density is high enough the ohmic voltage drop will be pronounced. In this regime the current across the boundary will be drift - limited.

DISCUSSION

A 2D FEM simulation was performed with two different seeds at the cathode (cf. Fig. 4). As boundary condition a 40 ns triangular current sweep with a peak current of 60 µA was used. Figure 3 shows the simulated resistance time characteristic in which switching is evident.

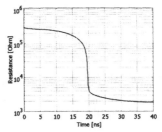

Figure 3: Simulated resistance time characteristic for a 2D FEM simulation of an ECM cell with two different seeds (cf. Figure 4) using a 40 ns triangular current sweep.

The time evolution of filamentary growth in the simulation with two different seeds is shown in Figure 4. Firstly, filaments start growing simultaneously at both seeds. As soon as the right filament approaches half of the length to the counter electrode, the growth speed of the other filament slows down and finally stops growing. The electrical field at the tip of the right filament is so high, that nearly the entire current flows through it and less through the other. This demonstrates that only one filament is responsible for set operation.

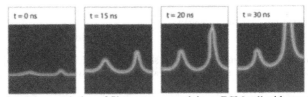

Figure 4: Simulated time evolution of filamentary growth in an ECM cell with two different seeds using a current sweep.

A 2D axial symmetric simulation was carried out and compared to a simulation with the resistor based 1D model. In both simulations a voltage sweep with a maximum voltage of 2 V and a rise time of 1.2 µs was used. The current compliance was set to 10.1µA. Since the shape of the filament and the conduction paths in the FEM model deviate from the shape in the resistor based model, the lateral areas of the filament and the solid electrolyte in the resistor based model were fitted such as the slope in the off resistance state coincide. As shown in Figure 5 the simulated I/V - curves match very well. Thus, the resistor based model appears to be a suitable simplification. Furthermore a good agreement to experiment is observed, comparing the simulation results with a measured I/V - curve (see Figure 6) of a WO$_3$ based ECM cell. The measurement was carried out using a voltage sweep with 10 µA current compliance.

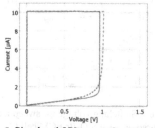

Figure 5: Simulated I/V - curve for 2D axial symmetric FEM model (solid line) and 1D model (dashed line) using a voltage sweep.

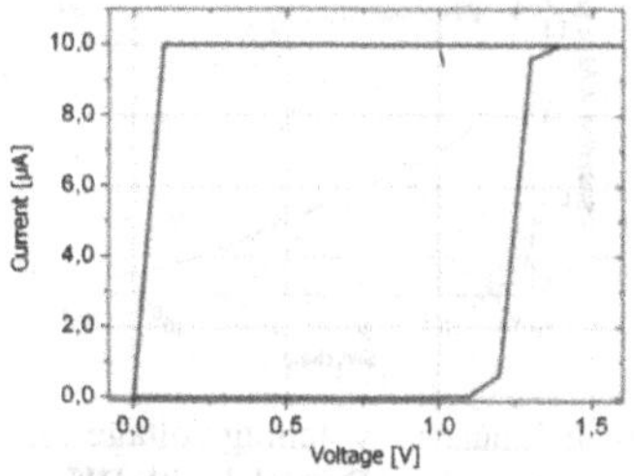

Figure 6: I/V –curve of a Cu/WO$_3$/Pt ECM cell using a triangular voltage sweep.

Figure 7 shows the simulated I/V characteristic for a current driven resistor based simulation of an ECM cell using BV boundary conditions. In this simulation a 260 ns current pulse with 1 µA maximum current was used. The nonlinearity in the off state due to BV boundary conditions can be clearly seen, resulting in a higher resistance at low voltages.

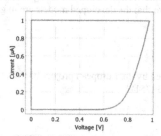

Figure 7: Simulated I/V – curve using the resistor based model with BV boundary conditions

Using the resistor based model with BV boundary conditions, a series of 100 single simulations with different current pulse amplitudes were carried out. The maximum reached voltage within each simulation was taken as switching voltage and the time when the filament reaches the counter electrode as switching time. The resulting voltage time dependence is shown in Figure 8. Three different regimes can be discriminated as predicted in theory. For low voltages and switching times above 1 s the linear electron transfer limited regime is visible. At faster switching times the relation between switching voltage and switching time becomes exponential, being in the electron transfer limited regime for higher overvoltages. For switching times faster than 100 ns the drift limited regime is apparent. A similar behavior was observed in experiment (see Figure 9, taken from [2]) for SiO_2 based ECM cells. In these experiments the sweep rate of a voltage sweep is varied (see inset of Fig. 7) and the set voltage is recorded. In this case the linear and exponential electron transfer limited regimes are visible. The transition between the two regimes occurs around a sweep rate of 20 mV/s.

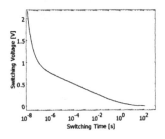
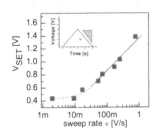

Figure 8: Simulated switching voltage vs. switching time for 1D model with BV boundary conditions.

Figure 9: Switching voltage vs. sweep rate measured on a Cu/SiO$_2$/Ir ECM cell [2]

CONCLUSIONS

In the present study a FEM simulation model and a resistor based 1D model were developed, capable of simulating the set operation in ECM cells. It could be demonstrated that just one filament is responsible for set operation. Simulations based on the resistor based 1D model with BV boundary conditions showed that switching time decreases with increasing switching voltage. It was shown that the developed models are in good agreement with experimental results.

ACKNOWLEDGMENTS

This work was supported by the European project EMMA "Emerging Materials for Mass storage Architectures" (FP6-033751).

REFERENCES

1. R. Waser and M. Aono, *Nature Materials*, USA, 6 (2007) 833-840
2. C. Schindler, G. Staikov, R. Waser, *Applied Physics Letters* 94, (2009) 72109/1-3
3. M. N. Kozicki, C. Gopalan, M. Balakrishnan and M. Mitkova, *IEEE Transactions on Nanotechnology*, USA, 5 (2006) 535-44
4. C. Kuegeler, C. Nauenheim, M. Meier, R. Ruediger and R. Waser, *Proceedings of Nonvolatile Memory Technology Symposium*, (2008) 59-63
5. J. A. Sethian, *Proc. Natl. Acad. Sci USA*, Vol.93, pp. 1591-1595 (1996)

Resistive Switching RAM II

Mater. Res. Soc. Symp. Proc. Vol. 1160 © 2009 Materials Research Society   1160-H10-01

Research Progress in the Resistance Switching of Transition Metal Oxides for RRAM Application: Switching Mechanism and Properties Optimization

Qun Wang, Xiaomin Li, Lidong Chen, Xun Cao, Rui Yang, and Weidong Yu

State Key Laboratory of High Performance Ceramics & Superfine Microstructure, Shanghai Institute of Ceramics, Chinese Academy of Sciences, 1295 Ding Xi Road, Shanghai 200050, People's Republic of China

ABSTRACT

Electric-induced resistance switching (EIRS) effect based on transition metal (TM) oxides, such as perovskite manganites ($Pr_{1-x}Ca_xMnO_3$, $La_{1-x}Ca_xMnO_3$) and binary oxides (NiO, TiO_2 and CoO) etc, has attracted great interest for potential applications in next generation nonvolatile memory known as resistance random access memory (RRAM). Compared with other nonvolatile memories, RRAM has several advantages, such as fast erasing times, high storage densities, and low operating consumption. Up to date, the switching mechanism, property improvement and new materials exploitation are still the hotspots in RRAM research.

In this report, the main results of resistance switching of two kinds of TM oxides including $La_{0.7}Ca_{0.3}MnO_3$ and TiO_2 were presented. Based on the I-V characteristics, the field-direction dependence of resistance switching (RS) behavior, and the conduction process analysis, the EIRS mechanisms were studied in detail. For the $La_{0.7}Ca_{0.3}MnO_3$ film, the EIRS mechanism was related to the carrier injected space charge limited current (SCLC) conduction controlled by the traps existing at the interface between top electrode and $La_{0.7}Ca_{0.3}MnO_3$ film. The RS behavior is produced by the trapping/detrapping process of carriers under different voltages. For the TiO_2 film, both unipolar and bipolar RS behavior can be obtained in our experiments. The interface controlled filamentary mechanism was proposed to explain the unipolar EIRS in nanocrystalline TiO_2 thin films, while the bipolar RS behavior may be related to the charge trapping or detrapping effect. In addition, it was confirmed that the I-V sweeps in vacuum environment, the applying of asymmetry pulse pairs and the oxygen annealing of films can improve the endurance of the EIRS devices.

INTRODUCTION

Electric-induced resistance switching (EIRS) based on transition metal (TM) oxides has recently attracted considerable research interest due to their potential application for resistance random access memory (RRAM), which is considered as the promising next-generation nonvolatile memory with some advantages of drastically reduced power consumption, fast switching speed and nondestructive readout [1-3]. Many models were proposed such as the modification of the Schottky barrier height by trapped charge carriers [4], pulse-generated crystalline defects [5], interface metal- insulator transition [6-7], trap-controlled space-charge-limited current (TC-SCLC) [8-10], the formation of a conductive filamentary path [11,12], the electrical field-induced migration of oxygen vacancies [13-15], and the polaron localization at the metal-oxide interfacial region [16], etc. However, all of the models leave unanswered questions. It seems that the EIRS is neither a pure interfacial nor a pure volume effect. Both the film and the electrode influence the resistance behavior under the applied voltage. EPIR memory is still at the stage of "proof-of-concept", much research is still required, from which the reversible transition mechanisms can be investigated deeply and the memory performance can be optimized.

Though full understanding of the EIRS effect is still lacking, charge trapping near the interface is thought to be at the core of the mechanism. Though there are some research works on this topic, much research is still required in revealing the storage origin, choosing the suitable materials and improving the memory feature. In this report, we will summarize the main results of the resistance switching investigation based on transition metal (TM) oxides, such as perovskite manganites ($La_{0.7}Ca_{0.3}MnO_3$) and binary oxides (TiO_2) etc, in our group in Shanghai Institute of Ceramics. It is demonstrated that two mechanisms dominate the EIRS behavior for the different transition metal oxides. And the endurance of the EIRS memory was improved by imposing the post-treatment and optimizing the pulse model.

EXPERIMENT

For $La_{0.7}Ca_{0.3}MnO_3$ film-based EIRS cell, the LCMO films of 200-nm-thick were deposited on the $Pt/Ti/SiO_2/Si$ substrates by the pulse laser deposition system. The oxygen partial pressure and substrate temperature during deposition were maintained at 0.02 Pa and 650 °C, respectively. The film deposition time is 120 min. Ag contact pads of radius $r \approx 0.3$ mm and a separation of 5 mm were painted on the top of the LCMO films.

For TiO_2 film-based $Pt/TiO_2/Pt$ and $Cu/TiO_2/Pt$ cells, about 100 nm thick TiO_2 film was fabricated on Pt (111)/Ti/SiO_2/Si substrates by thermal oxidation of evaporated Ti films at 700 °C using a tube furnace. The Ti films were firstly grown on Pt (111)/Ti/SiO_2/Si substrates by electron beam evaporation at room temperature. During the Ti metal evaporation, the pressure inside the chamber was less 1.3×10^{-4} Pa. The oxygen flow rate was fixed at about 30 cc / min. The tops Pt and Cu electrode (100 nm thick) with the protective Au layer (20 nm thick) of 100 μm in diameter were deposited by electron beam evaporation with a shadow mask at room temperature, respectively.

For the prepared EIRS cells, the electrical measurements were conducted by two-probe method using a dc source meter (Keithley 2410-c) and a pulse generator (Agilent 81104A). An electric pulse passing from the top electrode to the films is defined as the positive direction. The resistance was measured by the standard two-wire $I–V$ curve method using a 1μA dc current as the measuring current. The measuring current was applied in both positive and negative directions and the resistance was obtained from the average value in the two opposite directions. The $I-V$ characteristics were measured by voltage sweeping as $0 \rightarrow + V_{max} \rightarrow 0 \rightarrow -V_{max} \rightarrow 0$. All the measurement was performed at room temperature (RT).

RESULTS AND DISCUSSION

Storage properties of manganites ($La_{1-x}Ca_xMnO_3$) film based EIRS device

Figure 1 shows a schematic diagram of the manganites ($La_{1-x}Ca_xMnO_3$) film based EIRS

110

cell and the *I-V* characteristic of a Ag/La$_{0.7}$Ca$_{0.3}$MnO$_3$ /Pt heterostructure at room temperature under the voltage sweep 0 →2 V →0 →-2 V →0. Hysteresis and asymmetry phenomena were observed clearly. It can be seen that the resistance decreased under the positive voltage sweep region and increased under the negative voltage sweep region [9]. Figure 2 shows the two stable resistance states induced by applied pulse voltage with a good reproducible behavior.

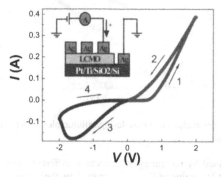

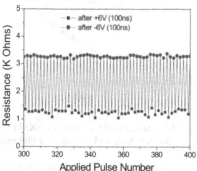

Fig.1 *I-V* characteristics of a Ag/LCMO/Pt heterostructure at voltage sweep 0→+2V→0→-2V→0. Arrows indicate sweeping directions. The inset shows the schematic of the samples and the measurement.

Fig. 2 The switching cycling characteristics of as-prepared Ag/LCMO/Pt heterostructure.

The stereo-type conduction mechanisms in semiconductors or insulators are grossly classified as the thermionic emission limited current, the Poole-Frenkel (PF) emission, the Fowler-Nordheim tunneling, and the space charge limited current (SCLC) mechanisms. These models can be distinguished via the isothermal *I-V* correlation [9]. Figures 3(a) and 3(b) show the logarithmic plots of the *I-V* curves for the positive and negative voltage sweep regions. In the 0 →+2 V region (Fig.3 (a)), *I-V* curve shows linear behavior under low voltage (<*Von* ≈0.13 V), and then quadratic. At the voltage V_T (~0.6 V), the current rises rapidly and the slope reaches 8~10. Then, the current rises slowly and the slope reduces to 2-3. The charge transport behavior in LCMO film can be explained by one-hole carrier injected trap-controlled space charge limited (SCL) conduction mechanism. When the applied voltage reaches *Von*, the carrier transport process changes from the Ohmic to SCL conduction. In this case, the carrier transit time is equal to the Ohmic relaxation time. When *V>Von*, the injected excess carriers dominate the thermally generated carriers since the injected carrier transit time is too short for their charges to be relaxed by the thermally generated carriers. Furthermore, the gradual increase (slope=8~10) in the trap-filled limit (TFL) region (V>0.6V) indicates the existence of traps distribution in traps-level energies in present experiment. The carrier transport mechanism experiences the process of Ohmic→ SCL conduction (controlled by single shallow trap) → TFL (controlled by traps with exponential distribution in energy) → SCL conduction (controlled by Child's law) → Ohmic in the region of positive voltage sweeping. In the negative bias region, as shown in Fig.3 (b), the carrier transport mechanism can be described as Ohmic→ SCL conduction (controlled by

111

Child's law) → SCL conduction (controlled by single shallow trap) → Ohmic.

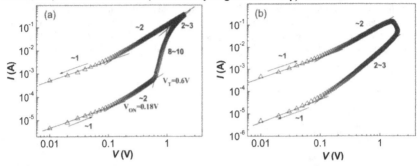

Fig. 3 *I-V* characteristics of a Ag/LCMO/Pt heterostructure in a double-logarithmic plot under (a) positive bias, and (b) negative bias regions.

I-V hysteresis behaviors were also examined by measuring *I-V* curves at different sweep ranges. In Figs. 4(a) and 4(b) it can seen that at low value of V_{max}, *I-V* curves show the transition from Ohmic to SCL conduction controlled by single shallow traps in positive bias region but no hysteresis, which indicates the trapped carriers can be released by traps with the removing of bias at this stage, and then the resistance change is unretentive. With raising V_{max}, *I-V* curves show TFL conduction (controlled by traps with exponential distribution in energy) in the positive sweep region, meanwhile the hysteresis can be observed, which indicates the injected carriers trapped with exponential distribution in energy remain after removing the positive bias until a negative threshold voltage of is applied. As the V_{max} rises to 2 V, *I-V* curves show the gradual transition from TFL to trap-free SCL conduction, where the hysteresis and resistance change approach to be saturated [9].

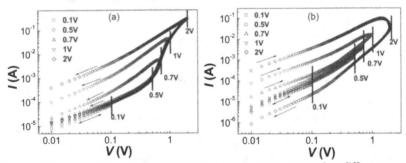

Fig. 4 *I-V* characteristics of a Ag/LCMO/Pt heterostructure measured at different ranges of voltage sweep under (a) positive, and (b) negative bias regions.

The above results indicate that the appearance of hysteresis depends on the carrier trapping levels. As the positive bias increased, the trapping level increases with increasing injected carrier density and the quasi-Fermi level (E_p) shifts towards the valence band. Only when the E_p intersects the exponential trap distribution with the increasing bias, the retention of trapped carrier can take the effect, leading to the I-V hysteresis and asymmetry of Ag/LCMO/Pt heterostructures. The resistance switching induced by pulse voltage originates from the change of trap distribution in energy in the films. The resistance states are dependent on the carrier trapping levels in the films, namely the high trapping level corresponding to low resistance state (LRS) while the low one corresponding to high resistance state (HRS).

Storage properties of binary transition-metal oxide (TiO₂) film based EIRS device

The Pt/TiO₂/Pt heterostructure was prepared to study the unipolar resistance switching behaviors. Rutile TiO₂ thin films were prepared by thermal oxidation of Ti films, which was demonstrated by XRD and Raman spectra. Figure 5 (a) shows the typical I-V curves obtained during the resistance switching of TiO₂ films. The switching from the LRS to the HRS is called reset process, and the switching from HRS to LRS is called set process. During the set process, the current compliance is applied in order to avoid permanent damage to the samples. Figure 5 (b) shows the resistance switching cycling characteristics of Pt/TiO₂/Pt heterostructure.

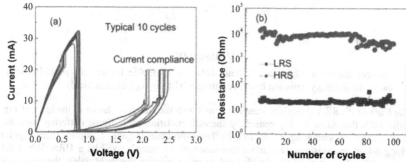

Fig. 5. (a) Typical I-V curves obtained at compliance currents of 20 mA during reset and set processes of Pt/TiO₂/Pt heterostructure, and (b) the HRS and LRS with 100 switching cycles in the unipolar resistive switching.

In order to explore the origin of the switching characteristics of TiO₂ films, the conduction mechanisms of the HRS and LRS were investigated. Here, typical switching cycles were selected, and the relationship of current versus voltage (I-V) was redrawn in double logarithmic plot, as shown in Fig. 6. Because the slope in low field is close to 1, the conduction mechanisms of both LRS and HRS in low-field region are believed to be an Ohmic behavior. At higher electric field, the conduction mechanism of LRS still obeys the Ohmic behavior, but a non-linear I-V characteristic appears at higher electric field for HRS. The classical non-linear conduction mechanisms including Schottky emission, Poole-Frenkel emission, and space charge limited current are adopted to fit the nonlinearity of the measured I-V relation. It is found that the Schottky emission has the best fit. The linear dependence of $\ln I$ versus $V^{1/2}$ is shown in the inset of Fig. 6, which indicates the typical Schottky emission behavior.

113

The mechanisms for unipolar resistive switching of the transition metal oxides are interesting but still controversial. Among many proposed models, the filamentary mechanism is well accepted by most researchers so far. This is in agreement with our experiment result that the conduction mechanism of LRS obeys the Ohmic behavior, but that of HRS obeys the Schottky-like barrier model, which indicates that the formation/rupture of conductive filaments is referred to explain the unipolar RS effect [17-19]. It is assumed that the filaments are formed by percolation of some kind of defects. In TiO_2, the most probable defects are titanium interstitials or oxygen vacancies. The migration of defects under the applied bias voltage is considered for inducing the formation and rupture of the filaments in TiO_2 thin films [20,21].

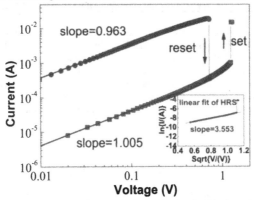

Fig. 6. The conducting mechanism is Ohmic behavior in double logarithmic plot. The inset shows the Schottky emission by curve fitting of HRS in high electric field.

The $Cu/TiO_2/Pt$ RRAM devices were fabricated with the structure shown in the inset of Fig. 7. Rutile TiO_2 thin films were prepared by thermal oxidation of Ti films. Differing from the unipolar resistance switching in $Pt/TiO_2/Pt$ heterostructure, figure 7 shows their bipolar switching I-V characteristics. The initial resistance of the sample exhibits a HRS. When the voltage is swept from zero to the positive values and reaches a threshold value, the resistance begins to decreases (defined as the set process). When the voltage is subsequently swept from the positive to the negative value, under a certain negative voltage, the resistance of the sample switches back from the low LRS to the HRS (defined as the reset process).

To confirm the EPIR characteristics, the switching operation in the pulse mode is shown in Fig. 8. A 50 ns voltage pulse of 8 V switches the device into the LRS, while a 0.5 μs voltage pulse of -5.3 V is used to turn the device state back into the HRS. From this result, it is clear that high speed bipolar switching in TiO_2 films is possible. After 10^4 pulse cycles, the resistance change ratio [defined as : $(R_{HRS}-R_{LRS})/R_{LRS}$] is still retained about 30%, which can meet the need of practical applications for RRAM.

There are many works to study the unipolar switching mechanism in binary transition-metal oxide film based EIRS device, which attribute it to filament conduction theory [17-21]. But the mechanism of bipolar resistance switching behavior in binary transition-metal oxide EIRS device is rarely to be reported. Dealing with the resistance change level in the RS process and the probable defects of titanium interstitials or oxygen vacancies in TiO_2, we suggest that this bipolar RS behavior may be related with the migration of defects under the applied bias voltage and the carriers trapping/detrapping effect at the interface. But the exact details of the actual mechanism of bipolar resistive switching in TiO_2 films are still under studies.

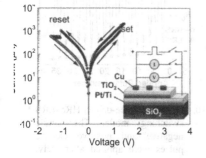

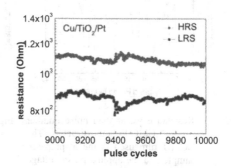

Fig.7. *I-V* characteristics of Cu/TiO$_2$/ Pt device (The inset shows the schematic EPIR device structure and electrical measurement setup).

Fig.8. The endurance characteristics of the bipolar resistive switching of the TiO$_2$ films.

A case study of endurance improvement of the EIRS devices

Experiments demonstrate that the optimized pulse applied modes have effects on improving the retention behavior of Ag/LCMO/Pt heterostructure. Figure 9 shows the relationship between the resistance and the applied pulse number when the pulse voltage was fixed at 5V for an as-prepared sample. In the first cycle, the resistance increased to about 150 kΩ by applying only one pulse. In order to maintain the resistance as high as the level of about 150 kΩ, three pulses were needed for the second circle. The pulse number needed for maintaining about 150 kΩ resistance increased with the circle number. It was 4, 6 and 7 for the third, fourth and fifth circles, respectively. The induced high resistance state is improved obviously under this asymmetric pulse model with more times of negative electric pulses [22].

In the single pulse-driven LRS-HRS switching mode, it is found that lower positive voltage and higher negative pulse voltage applied alternately have the noticeable effect on improving the HRS retention behavior. Figure 10 showed the EPIR ratio versus time for the Ag/La$_{0.7}$Ca$_{0.3}$MnO$_3$/Pt heterostructure after −14V and +8 V electric pulses were applied alternately. The high resistance states after several switching circles showed better retention behavior with time. In addition, the LRS retention behavior was also improved easily based on the reasonable pulse applied modes [23].

115

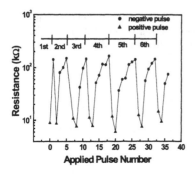

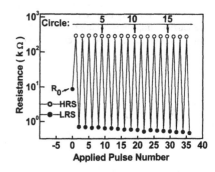

Fig. 9. Resistance vs. applied pulse number for Ag/LCMO/Pt heterostructure. The LCMO film used is the as-prepared sample and the applied pulse voltage is 5 V.

Fig. 10. The single pulse-driven HRS-LRS switching mode. The single negative and positive electric pulses were applied alternately. The negative pulse voltage was fixed at 14 V. The positive pulse voltage was fixed at 8 V.

Besides, it was found that the key point of the fatigue occurrence lies in the reaction between the electrode and the oxygen in air. Therefore, it is reasonable and practicable to improve the fatigue endurance by preventing the oxygen in the environment from the reduction-oxidation equilibrium before the fatigue takes place. It has been proved that the I-V sweeps in vacuum environment ($<$100 Pa) has a significant effect on improving the fatigue endurance of the resistance switching in Ag/LCMO/Pt heterostructures. As shown in Fig.11, the asymmetric fatigue can be refreshed by applying a voltage sweep in vacuum condition ($<$100 Pa). The fatigue and its refreshment are related to the increase and decrease in threshold voltage in the resistance switching, respectively. The results indicate that fatigue endurance in resistance switching of Ag/LCMO/Pt heterostructures can be improved through the operation of environmental oxygen pressure [24].

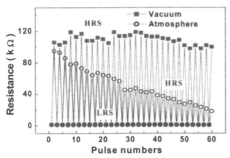

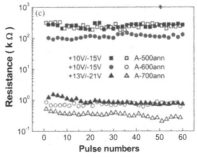

Fig. 11 Improvement of the fatigue by applying a voltage sweep in vacuum condition.

Fig. 12 Pulsed bias induced resistance switching of Ag-LCMO-Pt heterostructures with rapid post-annealing process of LCMO films in oxygen at different temperatures of 500°C, 600 °C and 700 °C. Each pulse pair was composed of one positive and one negative pulse.

In these experiments, it was also observed that the fatigue tendency seemed weaker for the annealed samples than for the as-prepared samples, i.e. the amplitude of the resistance decrease for the as-prepared samples decreased more rapidly than the annealed samples. Figure 12 shows resistance switching of Ag/LCMO/Pt heterostructure with annealed films at different temperatures in oxygen. Consistent with the I-V hysteresis, the A-600 (annealed at 600 °C) heterostructure exhibit much larger $\Delta R/R$ than A-700 (annealed at 700 °C). On the other hand, the resistance switching of A-700 needs larger voltage of pulsed bias. A-500 (annealed at 500 °C) heterostructure show no reversible resistance switching. The above results indicate that the LCMO films-based EIRS cell can be improved in fatigue property by post-annealing process in oxygen at temperatures of 600 °C through adjusting the crystallization of LCMO film [25].

CONCLUSIONS

Two kinds of the transition-metal oxides based EIRS cell were prepared on $Pt/Ti/SiO_2/Si$ substrates, and some conclusions can be drawn. For manganite film based EIRS devices, it has the best performance in endurance and stability properties, while it is hard to obtain the films with homogeneous composition. For binary transition-metal oxide (TiO_2) film based EIRS device, besides the available high resistance-switching ratio (unipolar RS) and high switching speed (bipolar RS), it has the advantages in preparation technology with low cost, but need to improve its stability and endurance. The storage mechanisms of two kinds EIRS devices were studied. For the $La_{0.7}Ca_{0.3}MnO_3$ thin film, the EIRS mechanism is related to the modulation of the space charge limited conduction at the interface between TE and the $La_{0.7}Ca_{0.3}MnO_3$ film. However, the interface controlled filamentary mechanism was suggested to be the dominant explanation for unipolar EIRS in nanocrystalline TiO_2 thin films. In addition, the I-V sweeps in vacuum environment, asymmetry pulse pairs and oxygen annealing of films can improve the endurance of the EIRS property. Our researches will provide some meaningful clues to understanding the EIRS mechanism and some useful pathways for the development of RRAM devices.

ACKNOWLEDGMENTS

This work was supported by the National Natural Science Foundation of China (Grant No. 50672116), the Hi-Tech Research and Development Program of China (Grant No 2006AA03Z308), Keystone Project of Shanghai Basic Research Program (08JC1420600),

Shanghai-AM Research and Development Fund (8700740900), and Nature Science Foundation of Shanghai (08ZR1421500).

REFERENCES

1. Yoshida C, Tsunoda K, Noshiro H and Sugiyama Y, *Appl. Phys. Lett.* **91**, 223510 (2007).
2. Baek I G, Lee M S, Seo S, Lee M J, Seo D H, Suh D -S, Park J C, Park S O, Kim H S, Yoo I K, Chung U-In and Moon J T, *Tech. Dig. IEDM* 587–90 (2004).
3. Zhuang W W, Pan W, Ulrich B D, Lee J J, Stecker L, Burmaster A, Evans D R, Hsu S T, Tajiri M, Shimaoka A, Inoue K, Naka T, Awaya N, Sakiyarma K, Wang Y, Liu S Q, Wu N J and Ignatiev A, *Tech. Dig. IEDM* 193–96 (2002).
4. A.Sawa, T.Fujii, M.Kawasaki, and Y. Tokura, *Appl. Phys. Lett.* **85**, 4073 (2004).
5. S.Tsui, A.Baikolov, J.Cmaidalka, Y.Y.Sun, Y.Q.Wang, Y.Y.Xue, C.W.Chu, L.Chen, and A.J.Jacobson, *Appl. Phys. Lett.* **85**, 317 (2004).
6. Rickard Fors, Sergey I.Khartsey, and Alexander M.Grishin, *Phys. Rev. B* **71**, 045305 (2005).
7. M.J.Rozenberg, I.H.Inoue, M.J.Sánchez, *Appl. Phys. Lett.* **88**, 033510 (2006).
8. A.Odagawa, H.Sato, I.H.Inoue, H.Akoh, M.Kawasaki, and Y.Tokura, *Phys. Rev. B* **70**, 224403 (2004).
9. D.S.Shang, Q.Wang, L.D.Chen, R.Dong, X.M.Li, and W.Q.Zhang, *Phys. Rev. B* **73**, 245427 (2006).
10. D.S.Shang, L. D. Chen, Q.Wang, W.Q. Zhang, Z.H.Wu, and X. M. Li, *Appl. Phys. Lett.* **89**, 172102 (2006).
11. S.Q.Liu, N.J.Wu, and A.Ignatiev, *Appl. Phys. Lett.* **76**, 2749 (2000).
12. I.H.Inoue, S.Yasuda, H.Akinaga, and H.Takagi, *Phys. Rev. B* **77**, 035105 (2008).
13. A.Baikalov, Y.Q.Wang, B.Shen, B.Lorenz, S.Tsui, Y.Y.Sun, Y.Y.Xue, and C.W.Chu, *Appl. Phys. Lett.* **83**, 957 (2003).
14. Krzysztof Szot, Wolfgang Speier, Gustav Bihlmayer and Rainer Waser, *Nature Mater.***5**, 312 (2006).
15. G.I.Meijer, *Science* **319**, 1625 (2008).
16. Ch. Jooss, J. Hoffmann, J. Fladerer, M. Ehrhardt, T. Beetz, L. Wu, and Y. Zhu, *Phys. Rev. B* **77**, 132409 (2008).
17. Yu Chao Yang, Feng Pan, Qi Liu, Ming Liu, and Fei Zeng, *Nano Letters*, March 10, 2009 DOI: 10.1021/n1900006g
18. B. J. Choi, D. S. Jeong, S. K. Kim, C. Rohde, S. Choi, J. H. Oh, H. J. Kim, C. S. Hwang, K. Szot, R. Waser, B. Reichenberg, and S. Tiedke, *J. Appl. Phys.* **98**, 033715(2005).
19. W. Y. Chang, Y. C. Lai, T. B. Wu, S. F. Wang, F. Chen, and M. J. Tsai, *Appl. Phys. Lett.* **92**, 022110 (2008).
20. S.C. Chae, J.S. Lee, S. Kim, S.B. Lee, S.H. Chang, C. Liu, B. Kahng, H. Shin, D.-W. Kim, C.U. Jung, S. Seo, M.-J. Lee, and T.W. Noh: *Adv. Mater.* **20**, 1154 (2008).
21. C. Rohde, B.J. Choi, D.S. Jeong, S. Choi, J.S. Zhao, and C.S. Hwang, *Appl. Phys. Lett.* **86**, 262907 (2005).
22. Rui Dong, Qun Wang, Lidong Chen, Tonglai Chen, Xiaomin Li, *Appl. Phys. A*, **80**, 13 (2005).
23. R. Dong, Q. Wang, L. D. Chen, D. S. Shang, T. L. Chen, X. M. Li, and W. Q. Zhang, *Appl. Phys. Lett.* **86**, 172107 (2005).

24. D S Shang, L D Chen, Q Wang, Z H Wu, W Q Zhang and X M Li, *J. Phys. D: Appl. Phys.* **40**, 5373 (2007).
25. D. S. Shang, L.D.Chen, Q.Wang, W.D.Yu, X.M.Li, J. R. Sun, and B. G. Shen, *J. Appl. Phys.* **105**, 063511 (2009).

Poster Session:
Resistive Switching RAM III

Mater. Res. Soc. Symp. Proc. Vol. 1160 © 2009 Materials Research Society   1160-H11-01

Investigation of resistance switching properties in undoped and Indium doped SrTiO₃ thin films prepared by pulsed laser deposition

Weidong Yu, Xiaomin Li, Yiwen Zhang, and Lidong Chen
State Key Laboratory of High Performance Ceramics & Superfine Microstructure,
Shanghai Institute of Ceramics, Chinese Academy of Sciences, 1295 Ding Xi Road,
Shanghai 200050, People's Republic of China

ABSTRACT

In this paper, the undoped $SrTiO_3$ (STO) and Indium doped STO ($SrTi_{1-x}In_xO_3$: STIO) thin films were grown on Pt/Ti/SiO₂/Si substrates by pulsed laser deposition with low substrate temperature. For undoped STO film, the influences of the forming processes on their resistive switching properties were studied by current and voltage controlled I-V sweeps, respectively. An obvious current controlled negative differential conductance phenomenon was found in both polarities of the electrical field. However, for low Indium doped STIO ($x=0.1$), the filament related resistance switching was observed in both the current and voltage I-V sweeps. And for high Indium doped STIO ($x=0.2$), a resistance switching with a reverse direction change to that in undoped STO can be obtained by a proper forming process. Based on these results, the reversible change of the Schottky like barrier at the grain boundary by the migration of oxygen vacancies were proposed to interpret the mechanism of the resistance switching.

INTRODUCTION

Some transition metal oxides including $SrTiO_3$, PCMO, NiO, CoO, and TiO_2 have attracted considerable attention as their stable reversible resistance change properties under a proper electrical exertion, which can be defined as "0" and "1" states for the potential application in a novel nonvolatile memories, named resistance random access memories (RRAM).[1-5] Among them, $SrTiO_3$ (STO) based materials, in particular $SrTiO_3$ based single crystal, which are well suited to serve the investigation of resistance reversible changes as it has a simple cubic structure in the room temperature, a model band insulator and can be modified easily by doping trivalent (Cr^{3+}) and pentavalent (Nb^{5+}) metal atoms. The filamentary mechanism related to the extended defects, such as dislocations, has been elucidated to explain the resistance switching of STO single crystals.[6] The electromigration of oxygen in the upper segment of the defect network drives the switching process. In addition, some reports exhibited the stable resistance switching properties in the heterojunctions of metal/STO [7] and STO/Nb-STO [8], which can also be explained by the roles of the interfaces of the junctions. However, due to the high preparation temperature (800°C) and strict substrate demand, the single crystal STO films can not be used for the Si integration devices at the present stage. The oxides used in IC technology such as SiO_2 and high K gate material HfO_2, ZrO_2 etc, are usually fabricated at low temperature (<500°C) with amorphous morphology to restrain the leakage current. So nanocrystalline STO thin film deposited at low temperature will adapt to the Si based IC process. On the other hand, STO ceramic is a conventional varistor material [9], which exhibits a symmetric nonlinear I-V characteristic. The high electrical nonlinearity is attributed to the formation of the electrical potential barrier at the grain boundaries, which imply that the grain boundaries will play an important effect on the resistance modulation of STO based materials. The similar results have been reported by Hirose in La doped STO ceramics chip [10]. The migration of oxygen vacancies near the grain boundaries under an electrical field was demonstrated as a main inducement for the resistance change in STO based materials. Although the ceramics can not be used for RRAM fabrication, the resistance switching originating from the grain boundaries will provide a new way to improve the stability and endurance of RRAM devices. In this paper, undoped and Indium $SrTiO_3$ thin films were grown on $Pt/Ti/SiO_2/Si$ substrates by pulsed laser deposition with low temperature and nanocrystalline structures. The resistive switching properties of the sandwich structures of Pt/STO/Pt and Pt/STIO/Pt were studied by current controlled and voltage controlled I-V sweeps. After a forming treatment of current controlled I-V sweeps, a stable resistance switching can be obtained in Pt/STO/Pt structure, meaning the resistance decreases in a positive sweep and abruptly increases in a negative sweep. In Pt/STIO/Pt structure with x=0.1, the filament related resistance switching was observed in both the current and voltage I-V sweeps. And in Pt/STIO/Pt structure with x=0.2, a resistance switching with a reverse direction change can be obtained. Based on these results, a possible mechanism of these effects was proposed and discussed.

EXPERIMENT

To investigate the resistance switching characteristics of the Pt/STO(STIO)/Pt sandwich structures, the STO and STIO ($SrTi_{1-x}In_xO_3$: x=0.1,0.2) films were grown on $Pt/Ti/SiO_2/Si(100)$ substrates by using pulsed laser deposition (PLD) employing a KrF excimer laser with wavelength of 248nm and pulse duration of 25ns. The targets were synthesized by a conventional solid state reaction and the nominal compositions were $SrTiO_3$, $SrTi_{0.9}In_{0.1}O_3$ and $SrTi_{0.8}In_{0.2}O_3$. The STO and STIO films were deposited at the substrate temperature of 400°C under oxygen pressure of 0.01Pa with constant laser energy density of $0.4J/cm^2$ and constant laser repetition of 5Hz. The distance between target and substrate was fixed to 7cm. The thickness of the STO film is around 100nm. Pt top electrodes with a diameter of ~100μm were prepared by electron beam evaporation under a base pressure of $< 5 \times 10^{-4}$ Pa with a dot-patterned metal shadow mask. The resistance and I-V characteristics were measured with the standard two-wire I-V curve method by precision Source Meter® (KEITHLEY 2410-C). The voltage bias was scanned with compliance current of 1mA and the maximum current in current controlled I-V sweep was 8mA. Here, the positive bias was defined as the current flowing from bottom Pt electrode to top Pt electrode. All measurements were performed at room temperature.

DISCUSSION

Figure 1 shows the I-V characteristics of Pt/STO/Pt multilayer units with the compliance currents (I_c) of 5mA. The forming sweeps have been performed by current controlled modes to obtain the stable resistance switching. It is worthy to note that in the case of voltage controlled model, the unit resistance decreases from the original value to a low resistance level and no resistance reversible change was found [11]. However, for current controlled model (figure 1a), there is two voltage decreases in the regions from 1.5mA to 3mA and -0.35mA to -0.4mA. Based on the definition of negative differential conductance (NDC) type [12], the configuration of current-voltage curve with the abscissa of voltage was called S type NDC that occurs in current controlled NDC model. In positive S-NDC region, the unit resistance decreases from the original value of 10kohm to the final value of 2kohm and the corresponding resistance change ratio reaches 5. Whereas, during negative sweep, a S-NDC process further induce the decrease of the unit resistance to 1.5kohm, and a voltage leap happens at the current of around 4.5mA then the incurred resistance change was recovered. The switching of the unit resistance can be realized by current controlled I-V sweeps. In order to elucidate the resistance change behavior in figure 1a, the corresponding voltage controlled I-V curve of figure 1a was shown together in figure 1b. By collating two I-V curves, it can be confirmed that the resistance decreases occurred during I-V sweeps are current controlled NDC (S-NDC) process, and the resistance increase is voltage controlled NDC (N-NDC) process. It is the first time that two type NDC processes were found in a system. It implies that the resistance increase and decrease of the STO film are induced by the different mechanism.

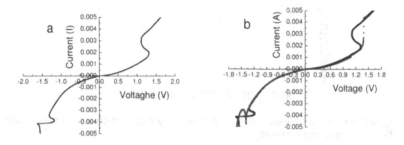

Figure 1 I-V characteristics of Pt/STO/Pt sandwich structure. (a) current controlled I-V sweep; (b) combined curves of voltage (red) and current (black) controlled I-V sweeps

Figure 2 shows the resistance switching characteristics in Pt/STO/Pt units with different I_c. It is found that the resistance of the unit decreases with the increase of the I_c in positive sweep. The resistance was degraded by the I-V sweep with increasing sweep voltage or current derived from the migration of the O vacancies.[13] When the I_c is less than 3mA, no resistance recover was happened in the negative sweeps. In other words, the threshold I_c for resistance switching is larger than 3mA. Actually, there is an indistinguishable and unstable resistance changes can be observed in the initial period of forming process ($I_c = 3mA$). The ratios of resistance changes in the units with the I_c of 5mA and 8mA are around 8 and 100, respectively. The high resistance ratio in the 8mA sweep induced not only by the decrease of the resistance of LRS but also the increase of the resistance of HRS. The resistances of HRS in units by sweeping at the I_c of 5mA and 8mA are comparable with the resistances at I_c of 3mA and 1mA, respectively. So we can suggest that it can be realized that the multi level resistance switching occurs at either HRS or LRS.

The resistance switching characteristics of STIO films are shown in Figure 3. In the case of indium substituted content (x) of 0.1, the original resistance of the sandwich structure reaches 10 Mohm, and a unipolar resistance switching (as shown in figure 3a) happened after the forming processes of both voltage and current controlled I-V sweeps. This switching behavior has been confirmed to be induced by a filamentary model, as that in binary metal oxides such as NiO, TiOx etc. However, for STIO with x=0.2, the original resistance of the sandwich structure increase to 100 Mohm. A resistance switching with a reverse change direction to that in undoped STO can be obtained by a careful forming process, as shown in figure 3b. In the negative S-NDC region, the unit resistance decreases from the original value of 500kohm to the final value of 0.5kohm and the corresponding resistance change ratio reaches 1000.

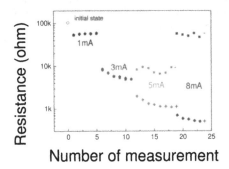

Figure 2 resistance changes of Pt/STO/Pt units by a current controlled sweep with different compliance currents

In order to understand the resistance switching mechanism of polycrystalline STO based materials, it is essential to probe the reason that the current controlled I-V sweep can induce the stable resistance switching. It is well known that the electrical nonlinearity of STO material is attributed to the formation of the electrical potential barrier at the grain boundaries. The depression and recovery of double Schottky-like barrier at STO grain boundary induce the decrease and recovery of the impedance of the STO based materials. No polarity can be found in this kind impedance change. In our case, during voltage controlled I-V sweep, the high electrical nonlinearity was found in positive sweep. The degradation of the unit resistance is induced by the oxygen vacancies migration from the top electrode to bottom electrode[13]. In this process, the barrier opposite to the electrical field was broken down by the continuous increase of the voltage. However, in the case of current controlled I-V sweep, the voltage may be decreased with the increase of the current since the barrier at the grain boundary decreases. The controllable change of the barrier improves the stability of the iterative migration of oxygen vacancies around the grain boundaries. It also prevents the oxygen vacancies from traversing grain boundary, which will induce the permanent degradation of the unit resistance.

It is to be noted that the doping of indium to STO matrix can be used to change the type of the majority carrier in STO based materials by substituting Ti^{4+} ions by In^{3+} ions [14]. The STO grain in undoped STO shows electronic conduction characteristics due to the existence of oxygen vacancies. The doping of In into STO compensates the function of oxygen vacancies. So, for STIO with x=0.1, the original resistance increase obviously. The Schottky like barrier at grain boundary is eliminated. The intrinsic switching behavior like that in STO crystal was disclosed. With the increase of In, the conduction type of STO translates to hole conduction characteristics. The effect of oxygen vacancies migration on the Schottky like barrier will be opposite to that in undoped STO. So, a reverse resistance switching was obtained.

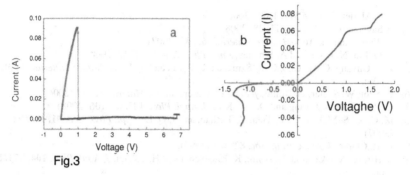

Fig.3

Figure 3 I-V characteristics of Pt/STIO/Pt units by current controlled sweep.
(a) x=0.1 (b) x=0.2

CONCLUSIONS

The undoped $SrTiO_3$ (STO) and Indium doped STO ($SrTi_{1-x}In_xO_3$: STIO) thin films were deposited on Pt bottom electrode substrates by pulsed laser deposition. Due to the intrinsic nonlinearity of I-V behaviors in undoped STO based materials, a reversible resistance change was observed only in current controlled sweep model. And the unit decreases from HRS to LRS by positive S-NDC process, and increases from LRS to HRS by negative N-NDC process. It can be realized that the multi level resistance switching occurs at either HRS or LRS by controlling the compliance current. The possible mechanism of this effect is attributed to the reversible resistance changes in Schottky like barrier at the grain boundaries at different level of current limitation. For low Indium doped STIO (x=0.1), the filament related resistance switching was observed in both the current and voltage I-V sweeps. And for high Indium doped STIO (x=0.2), a resistance switching with a reverse direction change to that in undoped STO can be obtained by a proper forming process. Based on these results, the reversible change of the Schottky like barrier at the grain boundary by the migration of oxygen vacancies were proposed to interpret the mechanism of the resistance switching. With the doping of In, the conduction type of STO translates from electronic conduction to hole conduction characteristics. The effect of oxygen vacancies migration on the Schottky like barrier in STIO becomes opposite to that in undoped STO.

ACKNOWLEDGMENTS

This work was sponsored by the Ministry of Science and Technology of China through the Hi-Tech Research and Development program of China(Grant no. 2006AA03Z308), National Nature Science Foundation of China(50672116), Keystone Project of Shanghai Basic Research

Program(08JC1420600), Shanghai-AM Research and Development Fund(8700740900) and Nature Science Foundation of Shanghai, (08ZR1421500).

REFERENCES

1. G. I. Meijer, *Science* **319**, 1625 (2008).
2. A.Sawa, *Materials today* **11**, 28 (2008).
3. R.Waser, and M. Aono, *Nature Materials* **6**, 833 (2007).
4. S. Q. Liu, N. J. Wu and A. Ignatiev, *Appl. Phys. Lett.* **76**, 2749 (2000).
5. J. Y. Ouyang, C. W. Chu, C. R. Szmanda, L. P. Ma and Y. Yang, *Nature Material* **3**, 4918 (2000).
6. K. Szot, W. Speier, G. Bihlmayer and R. Waser, *Nature Material* **5**, 312 (2006).
7. C. Park, Y. Seo, J. Jung and D. –W. Kim, *J. Appl. Phys.* **103**, 054106 (2008).
8. M. C. Ni, S. M. Guo, H. F. Tian, Y.G. Zhao, and J. Q. Li, *Appl. Phys. Lett.* **91**, 183502 (2007).
9. D. R. Clarke, *J. Am. Ceram. Soc.* **82**, 485 (1999).
10. S. Hirose, A. Nkayama, H. Niimi, K. Kageyama and H. Takagi, *J. Appl. Phys.* **104**, 053712 (2008).
11. Weidong Yu, Xiaomin Li, Yiwen Zhang, Rui Yang, Qun Wang, and Lidong Chen, *Adv. Mater. Res.* **66**, 199(2009).
12. G. Dearnaley, A.M. Stoneham and D.V.Morgan, *Rep. Prog. Phys.* **33**, 1129 (1970).
13. T. Waser, R. Baiatu and K.H.Hardtl, *J.Am.Ceram. Soc.* **73**, 1663 (1990).
14. Shouyu Dai, Huibin Lu, Fan Chen, Zhenghao Chen, Z. Y. Ren and D. H. L. Ng, *Appl. Phys. Lett.* **80**, 3545 (2002).

Mater. Res. Soc. Symp. Proc. Vol. 1160 © 2009 Materials Research Society   1160-H11-11

Lattice and electronic effects in rutile TiO$_2$ containing charged oxygen defects from *ab initio* calculations

Seong-Geon Park, Blanka Magyari-Köpe and Yoshio Nishi
Department of Materials Science and Electrical Engineering, Stanford University, Stanford, CA 94305, U.S.A.

ABSTRACT

We performed first-principle simulation for the study of oxygen vacancy defect in rutile TiO$_2$ based on density functional theory. The effects of a vacancy on the electronic structure of rutile TiO$_2$ were studied. Here we have employed neutral and charged oxygen vacancy in the supercell to address the resistance switching mechanism. Neutral vacancy induces the band gap states at deep level, ~0.7 eV below the conduction band minimum, which is occupied by highly localized electrons. The calculation results of positively charged oxygen vacancy show that larger atomic relaxation surrounding oxygen vacancy results in the stretching of Ti-O bond around vacancy, thus band gap states are formed near the conduction band minimum.

INTRODUCTION

ReRAM is very promising for advanced non-volatile memory technologies in terms of low operating power, high density, better non-volatility, fast switching speed, and compatibility with conventional CMOS process. There exist different switching characteristics that have been established in a variety of materials systems. In fact, it has been known that a few combinations of an oxide with metal electrodes can exhibit resistance switching characteristics. Recently different kind of experimental work has been reported to understand the resistance switching mechanism. Therefore several models have been proposed so far, such as charge trapping [1], the formation of conductive filament [2], the modulation of schottky barrier [3], and electro-chemical reduction and oxidation [4]. However, up to date none of these models is completely understood. In order to elucidate the operating principles accurately, in-depth understanding of the resistance switching mechanism at the atomistic level is necessary. In this study, we performed fist-principle simulation for the study of oxygen vacancy defect in rutile TiO$_2$, based on density functional theory (DFT). It is widely believed that defects play an important role in the switching of conducting state in transition metal oxide [5-7]. In fact, it is known that oxygen vacancies in TiO$_2$ generate excess electrons, previously occupying the oxygen 2p orbitals, thus n-type conducting behavior can be shown in oxygen deficient TiO$_2$. Here we calculated the lattice and electronic effects of oxygen vacancy in rutile TiO$_2$ using the local density approximation (LDA) with correction of on-site Coulomb interaction. However, it is found that the electronic structure of TiO$_2$ is not fully described by the on-site Coulomb interaction between 3d electrons, which is critical for the study of defect states within the band gap. In this study, we employ Coulomb interaction of both d and p character for the better description of electronic structure of TiO$_2$.

COMPUTATIONAL DETAILS

In this work we performed first-principles calculation for the study of oxygen vacancy in rutile TiO_2. Our calculations employ density functional theory (DFT) with plane waves as implemented in the Vienna *ab initio* simulation package (VASP) code [8]. The plane wave kinetic cutoff energy of 353 eV is used for expanding the electron wave functions. The Brillouin-zone (BZ) integration is performed using 4 x 4 x 4 Monkhorst-Pack grids in the first Brillouin zone of the supercell. The valence electron configurations are $3s^23p^63d^24s^2$ and $2s^22p^4$ for Ti and O atoms, respectively. The atoms are relaxed until the force on each atom is reduced to within 0.001 eV/Å in order to set up the equilibrium lattice parameter and the volume. The local density approximation with the correction of on-site Coulomb interactions (i.e., LDA+U) is used for the approximation of the exchange and correlation energies of electrons. The oxygen vacancy was introduced in a 2 x 2 x 3 supercell of rutile TiO_2.

RESULTS AND DISCUSSION

Fig. 1 shows the spin-density of states of rutile TiO_2 supercell with one neutral oxygen vacancy. We obtained energy band gap of 2.99 eV which is in good agreement with the experimental value of 3.0 eV. The presence of oxygen vacancy results in band gap states within the energy band gap. Fig 1 (b) and (c) is local density of states of one of nearest neighboring Ti atoms surrounding oxygen vacancy, which d and p orbital is plotted separately. It shows that band gap state has both d and p orbital character. This band gap state is fully occupied by two electrons that were previously hybridized with O $2p$ orbitals before the removal of oxygen atom. Our calculation shows that those electrons are strongly localized at the vicinity of oxygen vacancy site. Fig. 2 shows the electron localization function of TiO_2 with one oxygen vacancy which exhibits the degree of localization of electrons. It shows that there are highly localized d electrons near the oxygen vacancy site. In addition, the band gap state level is located deep in the band gap, at ~0.7 eV below the conduction band minimum, indicating that oxygen vacancy is not likely to provide electrons to the conduction band at room temperature. However it is well known that positively charged oxygen vacancy effectively dope TiO_2 with electrons, resulting in *n*-type semiconducting behavior [9]. The calculated spin-density of state and energy band structure of rutile TiO_2 with positively charged oxygen vacancy are shown in Fig. 3. As shown

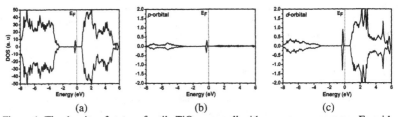

(a)   (b)   (c)

Figure 1. The density of states of rutile TiO_2 supercell with one oxygen vacancy. Fermi level is set to zero. (a) total density of states of supercell, (b) p-orbital local density of states, and (c) d-orbital local density of states of Ti atom adjacent to oxygen vacancy.

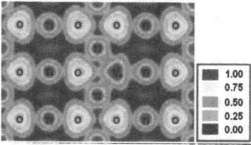

■	1.00
	0.75
■	0.50
	0.25
■	0.00

Figure 2. The electron localization function of (110) plane in rutile TiO_2. (1.0 : perfect localization, 0.5 : electron gas, 0~0.5 : low electron density)

in Fig. 1, both defect states are located at deep level and occupied by electrons for neutral oxygen vacancy. However, for singly charged oxygen vacancy, unoccupied spin-down defect state moves to the higher energy state near the conduction band, whereas spin-up defect state which is occupied slightly shifts down toward the valence band. In case of doubly charged oxygen vacancy, both defect states are observed near the conduction band minimum.

The difference in nature of the electronic state of oxygen vacancy in different charge states is linked to the different atomic arrangement around oxygen vacancy. The amount of the atomic relaxation in (110) plane is shown in Fig. 4 with respect to the unrelaxed vacancy configuration and as a function of the charge state of oxygen vacancy. Positive sign indicates outward relaxation. For neutral oxygen vacancy, three nearest neighboring Ti atoms are slightly relaxed outwardly by 0.3% of the equilibrium Ti-O bond length. This small displacement

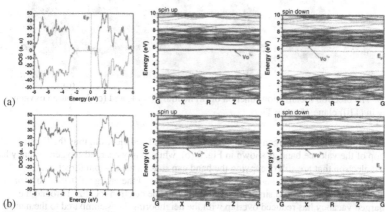

Figure 3. Spin-density of states and energy band structure of rutile TiO_2 with (a) 1+ charged oxygen vacancy, and (c) 2+ charged oxygen vacancy. The thick red line in energy band structure indicates the position of the band gap state level.

131

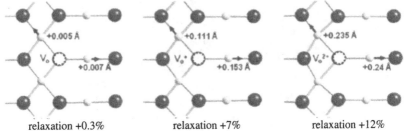

| relaxation +0.3% | relaxation +7% | relaxation +12% |

Figure 4. The local atomic relaxations around oxygen vacancy in the 0, 1+, 2+ charge states on (110) plane of rutile TiO_2.

indicates that most of the energy relaxation is conducted by electronic charge redistribution rather than the atomic relaxation. However these Ti atoms show much larger outward relaxation of 7% for singly charged vacancy and 12% for doubly charged vacancy, respectively. There were similar findings from Janotti et al. for ZnO [10]. They discussed about the origin of the large lattice relaxation and conclude that gain in electronic energy can be compensated for the strain energy to stretch the bond surrounding oxygen vacancy. In TiO_2, in the positively charged oxygen vacancy configuration, strain energy is exceeding the electronic energy so that Ti atoms show much larger outward relaxation, resulting in the stretching of the bond surrounding vacancy. The change in electronic structure of charged vacancy is observed in the electron localization function. Fig. 5 shows the electron localization function of different charged oxygen

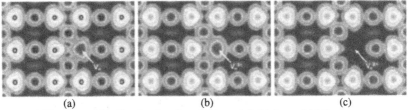

| (a) | (b) | (c) |

Figure 5. The electron localization function of (a) neutral, (b) 1+, and (c) 2+ charge oxygen vacancy in (110) plane of rutile TiO_2.

vacancy. In doubly charged oxygen vacancy, there is no localized electron, thus the Fermi level is at the top of the valence band as shown in Fig. 3 (b), while singly charged oxygen vacancy has a localized electron that are still occupying the band gap state in the up-spin direction. In addition, one should notice that there is a change in the shape of the charge distribution around the vacancy between neutral and singly charged vacancy. The electron localized around the singly charged vacancy has reduced overlap with the neighboring Ti as compared to the neutral oxygen vacancy, suggesting that the Coulomb interactions are weakened and Ti-O bonds are lengthened. The calculated Ewald energy was reduced by 6.92 eV/unit cell and 12.58 eV/unit cell by introducing singly and doubly charged vacancy, respectively. Consequently the removal of electrons gives rise to weaker Coulomb interactions around oxygen vacancy, thus the larger amount of relaxation is observed around the charged oxygen vacancy. Accordingly the different positions of band gap states from different charge state of oxygen vacancy shown in Fig. 3 are related to these equilibrium atomic configurations, which have different bond lengths surrounding oxygen vacancy.

In order to understand the resistance switching in ReRAM, on the other hand, it is necessary to consider the formation of conductive filament which is one of the candidates for the resistance switching mechanism. It was found that some kind of conductive path with high concentration of oxygen vacancies is formed between top and bottom electrode when the system reaches low resistance state [11]. In such case, as the distance between vacancies decreases, there will be strong interaction of vacancies resulting in the overlapping of electron wave function. Band gap states generated by charged oxygen vacancies may provide energy levels for electron hopping and therefore further studies are necessary to elucidate a possible mechanism of explaining the low and high resistance states of ReRAM.

CONCLUSIONS

In this paper, the lattice and electronic properties of rutile TiO_2 with oxygen vacancies were investigated to address the resistance switching mechanism in ReRAM. The calculations show that positively charged oxygen vacancy induces the Fermi level shifts and large relaxation of Ti atoms surrounding oxygen vacancy. The largest atomic relaxation surrounding doubly charged oxygen vacancy leads to the bond stretching surrounding oxygen vacancy, resulting in the formation of defect states near the conduction band minimum. These band gap states might be responsible for the electron doping and the low resistance state of rutile TiO_2 in ReRAM. We will conduct further studies to address this question.

ACKNOWLEDGMENTS

This work was supported by Stanford NMTRI project. The computational work was carried out through the National Nanotechnology Infrastructure Network's Computational drive (NNIN/C) at Harvard University.

REFERENCES

1. A. Chen, S. Haddad, Y. C. Wu, Z. Lan, T. N. Fang, and S. Kaza, Appl. Phys. Lett., vol. 91, pp. 123517, Sep 2007.
2. D. C. Kim, S. Seo, S. E. Ahn, D. S. Suh, M. J. Lee, B. H. Park, I. K. Yoo, I. G. Baek, H. J. Kim, E. K. Yim, J. E. Lee, S. O. Park, H. S. Kim, U.-In. Chung, J. T. Moon, and B. I. Ryu, Appl. Phys. Lett., vol. 88, pp. 202102, May 2006.
3. T. Fujii, M. Kawasaki, A. Sawa, H. Akoh, Y. Kawazoe, and Y. Tokura, Appl. Phys. Lett,. vol. 86, pp. 012107, Dec 2004.
4. X. Guo, C. Schindler, S. Menzel, and R. Waser, Appl. Phys. Lett., vol. 91, pp. 133513, Sep 2007.

133

5. M. Janousch, G. I. Meijer. U. Staub, B. Delly, S. F. Karg, and B. P. Andreasson, Adv. Mater., vol. 19, pp. 2232-2235, 2007.
6. H. Sim, D.-J. Seong, M. Chang, and H. Hwang, IEEE NVSMW. 21st., pp. 88-89, 2006.
7. A. Ignatiev, N. J. Wu, S. Q. Liu, X. Chen, Y. B. Nian, C. Papaginanni, J. Strozier, and Z. W Xing, NVMTS. 7th Annual, pp. 100-103, Nov 2006.
8. G. Kresse, and J. Hafiner, Phys. Rev. B., vol. 47, no. 1, pp. 558-561, Jan 1993.
9. E. G. Eror, J. Solid State Chem., vol. 38, pp. 281-87 (1981)
10. A. Janotti, and C. G. Van de Walle, Appl. Phys. Lett,. vol. 87, pp. 122102 (2005)
11. Janousch et al., Adv. Mat. 19, 2232 (2007)

Mater. Res. Soc. Symp. Proc. Vol. 1160 © 2009 Materials Research Society   1160-H11-13

Joe Sakai

Laboratoire LEMA, UMR 6157 CNRS/CEA, Université François Rabelais - Tours,

Parc de Grandmont 37200 Tours, France

ABSTRACT

The electric-field-induced resistance switching (EIRS) phenomenon on a VO_2 planar-type junction fabricated on a Al_2O_3 (0001) substrate was studied by performing current-voltage (I-V) measurement and optical microscope observation simultaneously. It was confirmed that current density J of the low-resistance-state (LRS) region is maintained constant at approximately 1.6×10^6 A/cm^2, while the volume of the LRS region was changed according to the current. A survey of the previous I-V traces on EIRS of VO_2 revealed that almost all the junctions so far had shown non-zero V-intercepts, which are attributed to the volume change of the LRS regions. The maintenance of high-J in the LRS region is considered to be related to the electrically-induced metallic phase mechanism reported in perovskite-type manganites.

INTRODUCTION

Electric-field-induced resistance switching (EIRS) of VO_2 was primarily reported by Bongers and Enz on 1966 [1]. This phenomenon, appearing as non-linear S-shape current-voltage (I-V) characteristics [2], has been widely researched up to now, raising interests in both pure and applied physics fields. The EIRS properties of VO_2 have lead researchers to propose some applications for switching devices [3,4] and oscillators [5]. In the first two decades, the sharp EIRS has been attributed to a thermally-induced structural phase transition from monoclinic to tetragonal structures [6], which is well-known to occur on VO_2 at 341 K [7]. Recently, some groups are trying to describe the EIRS of VO_2 by a Mott transition [3,8], or an avalanche breakdown [9] effects.

A series of I-V curves of VO_2 have often shown a peculiar feature that the asymptotic line in the low-resistance-state (LRS) region of the I-V trace does not pass the origin but crosses the voltage axis at non-zero value. In another expression, $dI / dV > I / V$ at LRS. Basically, the I-V line should be linear and pass the origin, thus it should be $dI / dV = I / V$, if the temperature and volume of the sample are unchanged. The reason of non-zero V-intercept has not been well discussed so far.

The present study focuses on the maintenance mechanism of the LRS region. Fortunately, the electric-field-induced LRS region of VO_2 is optically observable not only in IR but also in visible wavelength, and it should be quite informative to perform the electrical- and optical-observations simultaneously. Presenting the new results of electrical- and optical- observations, as well as surveying a series of previous results, I discuss several important questions on the maintenance of the LRS regions, such as why the V-intercepts of the asymptotic lines of LRS are non-zero in many cases, and what is occurring in the LRS regions.

EXPERIMENTAL SETUP

VO$_2$ thin film planar-type junctions were fabricated on Al$_2$O$_3$ (0001) single crystal substrates. The detailed processes of deposition of VO$_2$ thin films using a pulsed laser deposition method, and fabrication of junctions, are described elsewhere [9]. The VO$_2$ film for the present experiments is 200 nm in thickness. The dimension of the junction is 1500 μm in width and 10 μm in length. The experimental setup is schematically shown in figure 1. The triangle-shaped voltage wave was applied to the circuit. The important parameter in the observation of such a strongly non-linear sample is the series resistance R_s, which in the present case is the sum of two resistances, one resistor to monitor the circuit current and another resistor to suppress the maximum current. The optical microscope view of the junction was videotaped in parallel to the I-V measurements. All the experiments were performed in the air at room temperature (without temperature control).

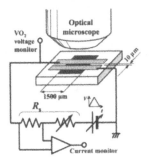

Figure 1. Schematic diagram of the experimental setup.

RESULTS

Figure 2(a) shows the I-V characteristics of the present VO$_2$ planar-type junction fabricated on a Al$_2$O$_3$ substrate. The sweeping time was 10 sec for one pair of increase and decrease runs. It is noteworthy that a shorter sweeping time of 1 sec resulted in the same I-V trace. A sharp resistance switching, similar to the previous reports, was observed. The asymptotic line of the LRS region crosses V-axis at $V = 0.6$ V (thus V-intercept is 0.6 V in this case).

It would be important to check whether the non-zero V-intercept feature is common to all the VO$_2$ materials so far, and how the V-intercepts and gradients at LRS are scattered among the samples. Figure 2(b) shows a series of I-V curves of VO$_2$ thin films previously reported. Only the reports that clearly show the behavior in their LRS regions were selected.

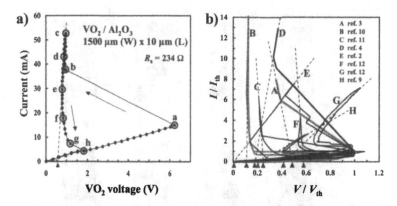

Figure 2. (a) *I-V* characteristics of a VO_2 planar-type junction. The labels a – h correspond to those on the optical images in figure 3. (b) A series of *I-V* curves of VO_2 thin films previously reported. All the voltage and current values are normalized by the voltage and current at the transition point in the increasing run. The voltage always implies the VO_2 sample voltage (If the source voltage has been plotted in the original figure, the values were transformed to sample voltage considering R_s). The dashed lines and filled triangles indicate the asymptotic lines in the LRS regions of the *I-V* traces and their *V*-intercepts, respectively, both in panels (a) and (b).

From the plot one can recognize that the behavior of LRS region strongly depends on the junctions. There exists a wide scattering among their *V*-intercepts and dI / dV values. Especially, it should be noted that trace F shows almost vertical increasing of current whereas trace G has a much smaller gradient at their LRS regions, in spite of the fact that these two junctions are on a single chip. On the other hand, it was confirmed that all the junctions have shown a non-zero *V*-intercept feature, or the relationship $dI / dV > I / V$, except junction E which has been used for the first *I-V* report of VO_2 on 1967 [2].

Figure 3 shows a series of optical microscope images that were taken simultaneously with the electric *I-V* trace. The dark-colored region appears and disappears at the very instances that the circuit current jumps up to the LRS and drops down to the high-resistance state (HRS), respectively. Therefore, there is no doubt that the dark region corresponds to the conductive LRS region. We have performed the comparison of the colors of VO_2 films of electric-field-induced LRS and thermally-induced LRS, and confirmed that the color in both states is identical, suggesting the tetragonal phase crystal structure of the electric-field-induced LRS [9]. In figure 3, one can also confirm that the width of the LRS region increases as the current increases.

The relationship between the current through the LRS region and its cross section is summarized in figure 4. LRS current values are calculated using the total junction current, HRS region width, junction voltage, and resistance at HRS. The cross section is the product of the LRS region width and the film thickness. Some of the previous data that mention both LRS region width and corresponding LRS current values are also plotted. This result clearly shows that the current density within the LRS region is as huge as approximately 1.6×10^6 A/cm^2.

137

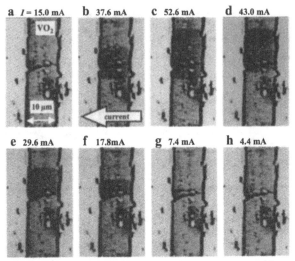

a $I = 15.0$ mA **b** 37.6 mA **c** 52.6 mA **d** 43.0 mA

VO$_2$

10 μm current

e 29.6 mA **f** 17.8mA **g** 7.4 mA **h** 4.4 mA

Figure 3. Optical microscope images of the same VO$_2$ planar junction as in figure 2(a). The labels a – h correspond to those on the points in figure 2(a).

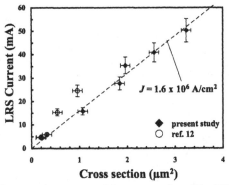

$J = 1.6 \times 10^6$ A/cm^2

♦ present study
○ ref. 12

Cross section (μm^2)

Figure 4. Relationship between the current and the cross section of the LRS region. Some data from a previous report are also plotted.

DISCUSSION

The volume of LRS increases according to the current (figure 3), and obviously, this is the direct reason why the V-intercept is non-zero in the present VO$_2$ junction. As a typical characteristic, there often exist the cases that the current increases where the voltage is unchanged [See traces B, C, and F in figure 2(b)]. Supposing the uniform current in the direction of thickness, this behavior suggests a feature that the current density J in the LRS region tends to be constant. In other words, the width of LRS region increases proportionally to the current.

The question at the moment is why J in the LRS tends to be always constant, both in the current increasing and decreasing runs. J should normally increase and decrease as the current does. Indeed, this is the case in VO_2 at temperatures higher than 341 K, where the sample is tetragonal-structured uniformly. In the case of VO_2 at field-induced LRS, the LRS region is narrow and therefore J is huge from the beginning, and it is kept huge even in the current decreasing run, as if J at the LRS cannot be larger or smaller than 1.6×10^6 A/cm^2 in this junction.

Supposing if this value is the upper limit of J in the LRS of VO_2, an injection of more electrons would force the LRS region to expand its territory, and vice versa. On the other hand, the picture that this J value is also the lower limit for LRS implies that this J is necessary in order to maintain the tetragonal structure of VO_2 at the electric-field-induced LRS. These observations strongly indicate that the electric-field-induced LRS is a nature that is controlled and maintained by an electric current of huge J. A model that the Joule heat plays the key role to maintain the LRS is not realistic because the I-V traces are unchanged under various sweeping speeds of source voltage. Note that it does not deny the possibility that (a part of) the LRS region is heated higher than 341 K.

Here, an important question arose, whether it is really possible for a current of 10^6 A/cm^2 to arrange the crystal structure from monoclinic to tetragonal without a support from the heat. To achieve it, the current flow should balance the force onto each vanadium ion which would suffer a repulsive force from another vanadium ion in the neighboring pair. Although it is not clear yet what mechanism would realize it, there is a report that a current of high density induces a metal phase of $Pr_{0.55}(Ca_{0.75}Sr_{0.25})_{0.45}MnO_3$ [13]. In this case the transformation from insulating to metallic phases is considered to be a "melting" of a charge-orbital-ordered state. Although the current density is 3×10^3 A/cm^2 which is a few order smaller than VO_2 case, the role that the current of huge density plays in maintaining LRS may be similar in both cases.

Coexistence of monoclinic and tetragonal regions with a quite sharp border suggests the competition of the two phases which are energetically separated by a potential barrier. This situation reminds me a double exchange mechanism in a series of perovskite-type manganites. In ferromagnetic $La_{1-x}Sr_xMnO_3$, for example, the localized spins at Mn ions are aligned in one direction, because the aligned spins would enhance the transfer integral of a conduction electrons and thus lower the total energy [14]. Analogizing this mechanism to the EIRS of VO_2, a picture can be drawn that the vanadium ions are aligned along c-axis because the aligned vanadium ions would make the transfer integral of the electrons higher and thus it is energetically gainful.

On the basis of the constant-J mechanism, it is rather ridiculous that there are still some junctions with small V-intercepts [D, E, and H in figure 2(b)]. Obviously the LRS of these junctions should be considered to behave in a manner different from the constant-J mechanism. Unfortunately, there has been no report on the optical observation of the LRS region of any small V-intercept junction. I consider it as a future subject.

The EIRS phenomenon on a VO_2 planar-type junction was studied by performing I-V measurements and optical microscope observations simultaneously. J of the LRS region is maintained constant at 1.6×10^6 A/cm^2, while the volume of the LRS region is changed according to the current. Almost all the VO_2 thin film junctions so far had shown non-zero V-intercepts, which can be attributed to the volume change of the LRS region. It is considered that the constant-J mechanism is related to the electrically-induced insulator-metal transition reported in perovskite-type manganites.

ACKNOWLEDGMENTS

The author is grateful to Prof. K. Okimura, Drs. A. Sawa, Y. Ogimoto, and J. Wolfman for their fruitful discussions. The present experiments were performed utilizing a setup in CERTeM consortium at Tours.

REFERENCES

1. P. F. Bongers and U. Enz, Philips Res. Rep. **21**, 387 (1966).
2. K. van Steensel, F. van de Burg, and C. Kooy, Philips Res. Rep. **22**, 170 (1967).
3. H. T. Kim, B. G. Chae, D. H. Youn, G. Kim, K. Y. Kang, S. J. Lee, K. Kim, and Y. S. Lim, Appl. Phys. Lett. **86**, 242101 (2005).
4. F. Dumas-Bouchiat, C. Champeaux, A. Catherinot, A. Crunteanu, and P. Blondy, Appl. Phys. Lett. **91**, 223505 (2007).
5. J. Sakai, J. Appl. Phys. **103**, 103708 (2008).
6. A. Mansingh and R. Singh, J. Phys. C **13**, 5725 (1980); and references therein.
7. F. J. Morin, Phys. Rev. Lett. **3**, 34 (1959).
8. G. Stefanovich, A. Pergament, and D. Stefanovich, J. Phys.: Condens. Matter **12**, 8837 (2000).
9. J. Sakai and M. Kurisu, Phys. Rev. B **78**, 033106 (2008).
10. C. N. Berglund and R. H. Walden, IEEE. Trans. Electron Devices **17**, 137 (1970).
11. J. C. Duchene, M. M. Terraillon, M. Pailly, and G. B. Adam, IEEE. Trans. Electron Devices **18**, 1151 (1971).
12. K. Okimura, N. Ezreena,Y. Sasakawa, and J. Sakai, Jpn. J. Appl. Phys. **48**, (2009) (in press).
13. N. Takubo and K. Miyano, Phys. Rev. B **76**, 184445 (2007).
14. A. Urushibara, Y. Moritomo, T. Arima, A. Asamitsu, G. Kido, and Y. Tokura, Phys. Rev. B **51**, 14103 (1995).

Poster Session:
Phase Change RAM I

Mater. Res. Soc. Symp. Proc. Vol. 1160 © 2009 Materials Research Society   1160-H12-05

Electrical resistance and structural changes on crystallization process of amorphous Ge-Te thin films

Yuta Saito, Yuji Sutou and Junichi Koike
Department of Materials Science, Tohoku University, 6-6-11-1016 Aoba-yama, Sendai 980-8579, Japan

ABSTRACT

The electrical resistance and structural changes on the crystallization process of sputter-deposited amorphous $Ge_{100-x}Te_x$ (x: 46-94) were investigated by two-point probe method. It was found that stoichiometric GeTe amorphous film crystallizes into rhombohedral GeTe single phase, which leads to a large electrical resistance drop, while off-stoichiometric Ge-Te amorphous films show two-stage crystallization or phase separation via single crystalline phase. Especially, $Ge_{33}Te_{67}$ amorphous film crystallizes first into metastable $GeTe_2$ single phase and then decomposes into α-GeTe and Te two-phase with a large electrical resistance change. The first crystallization temperature strongly depends on the composition. The $Ge_{33}Te_{67}$ film shows the highest crystallization temperature and activation energy for the first crystallization in the film with Te-rich composition.

INTRODUCTION

Recently, phase change random access memory (PCRAM) is attracting considerable attention because of low production cost and high scalability. In the PCRAM, the programming of bits is accomplished by phase transition between a high-electrical resistivity amorphous and a low-electrical resistivity crystalline states in a phase change material, which is induced by Joule heating. Since the difference of the electrical resistance between the amorphous and the crystalline states is very large, the stored bits can be read easily.

$Ge_2Sb_2Te_5$ (GST) phase change material, which is widely used for an optical disk memory such as DVD-RAM, has been much considered for the PCRAM. The GST shows high crystallization rate and good reversibility between amorphous and crystalline phases [1]. However, a high reset current which causes a high-power consumption is necessary because of its high melting temperature (~ 632 °C). In order to reduce the power consumption, eutectic alloys such as Sb-Te [2] and Ge-Sb [3] are attracting great attention due to their low melting points.

Ge-Te alloys have been suggested as the phase change material for the optical disk by Chen et al. in 1986 [4]. Stoichiometric GeTe possesses a higher crystallization temperature than the GST and has a high speed crystallization of < 30 ns, although it shows high melting temperature of ~ 724 °C. The Ge-Te alloy has an eutectic point in the Te-rich portion of around 85 at.% Te [5]. Therefore, the melting point decreases with increasing Te content from 724 °C (GeTe) to 375 °C ($Ge_{15}Te_{85}$). Although the GeTe compound is known to show the phenomenon of electrical switching [6], limited work has been done in the eutectic type Ge-Te alloys focusing on the phase change material for the PCRAM. In this study, the electrical resistance and crystalline structural changes on crystallization process in the Ge-Te thin films were investigated.

EXPERIMENTAL PROCEDURES

$Ge_{100-x}Te_x$ (x : 46-94) films with 200 nm thickness were deposited on SiO_2 (20 nm)/Si substrates by co-sputtering of GeTe and Te targets. *In situ* electrical resistance measurements during heating were performed by two-point probe method at a heating rate of 2 to 50 °C/min. in Ar atmosphere. X-ray diffraction (XRD) analysis was performed for structural identification at room temperature. X-ray spectra were taken in the 2θ range of 10-60° using Cu-Kα with a scanning step of 0.02°. The microstructure was investigated by transmission electron microscopy (TEM). The TEM specimens were prepared by mechanical polishing and ion milling with Ar ions. The compositions of these films were analyzed by energy dispersive X-ray spectroscopy (EDS).

RESULTS AND DISCUSSION

A. Temperature dependence of electrical resistance

Figure 1 shows the temperature dependence of the electrical resistance of the $Ge_{100-x}Te_x$ (x = 46, 50, 67, 94) thin films. The data of the GST is also shown for comparison. All as-deposited films are confirmed to be amorphous by the XRD and the electron diffraction patterns, and therefore, show a high resistance value. Composition dependence is seen in the resistance change behaviors, such as the crystallization temperature and the difference of the electrical resistance between the amorphous and the crystalline phases. Figures 2(a)-(c) show the XRD patterns of the as-deposited films and the films heated up to given temperatures in (a) the $Ge_{50}Te_{50}$, (b) the $Ge_{54}Te_{46}$ and (c) the $Ge_{33}Te_{67}$ films, respectively.

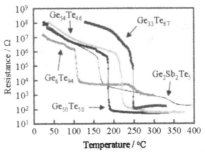

Figure 1. Temperature dependence of electrical resistance of $Ge_{100-x}Te_x$ and $Ge_2Sb_2Te_5$ thin films measured at heating rate of 8.7 °C/min.

(i) $Ge_{50}Te_{50}$ film

It is seen from Fig.1 and Fig.2 (a) that the stoichiometric $Ge_{50}Te_{50}$ amorphous film shows drastic decrease in the electrical resistance with crystallization into α-GeTe single phase at 190 °C. The rhombohedral α-GeTe is the stable phase of the stoichiometric GeTe compound at low temperature [7]. The crystallization temperature of the $Ge_{50}Te_{50}$ is higher than that of the conventional GST and the resistance change with crystallization is also larger. Moreover, the temperature dependence of the electrical resistance of the α-GeTe crystalline state is small, while

144

the electrical resistance of the GST gradually decreases with increasing temperature after the crystallization at 160 °C.

(ii) *Ge₅₄Te₄₆ and Ge₆Te₉₄ films*

   The Ge-rich Ge₅₄Te₄₆ and the Te-rich Ge₆Te₉₄ amorphous films show two stage crystallization processes upon heating. The amorphous Ge₅₄Te₄₆ film crystallizes first into α-GeTe crystalline at 220 °C with drastic resistance change as shown in Fig.1 and Fig.2(b). It was confirmed from the TEM observation that the Ge₅₄Te₄₆ film contains a small amount of amorphous phase after the first crystallization. By further heating, the residual amorphous phase crystallizes into Ge crystalline accompanying a small resistance change at 270 °C and then the electrical resistance remains almost constant. The XRD and the TEM observations indicate that the amorphous Ge₆Te₉₄ film mostly crystallizes first into Te crystalline at 110 °C and then the residual amorphous crystallizes into α-GeTe crystalline at 250 °C. Figure 1 also indicates that the resistance of the Te + α-GeTe two-phase state is about ten times higher than that of the other compositions.

(iii) *Ge₃₃Te₆₇ film*

   The Ge₃₃Te₆₇ amorphous film exhibits the highest electrical resistance in all compositions. It is noted that the temperature dependence of the electrical resistance just before the drastic drop is different from that of the other films. Namely, the negative slope of the electrical resistance gradually increases by heating. The XRD patterns in Fig.2(c) shows that the Ge₃₃Te₆₇ amorphous phase is maintained up to 225 °C, and then, crystallizes into the GeTe₂ crystalline by heating to 235 °C without the drastic resistance drop. The GeTe₂ crystal has been known to be a metastable phase, and was reported to appear only in a thin-film form (film thickness < 6 μm) [8].

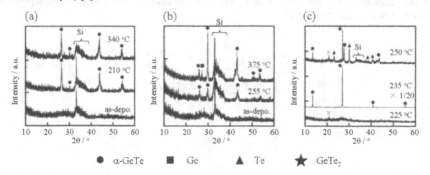

Figure 2. X-ray diffraction patterns of (a)GeTe, (b)Ge₅₄Te₄₆ and (c)Ge₃₃Te₆₇ films. Broad peak from 32° to 38° comes from Si substrate. The XRD pattern of (c)Ge₃₃Te₆₇ film heated to 235 °C is one-twentieth the intensity of the original one, because the meta-stable GeTe₂ compound has a strong (111) preferred orientation. The index × indicates the peak from the glue to fix the specimen.

   Figures 3(a) and (b) show a cross sectional TEM image and a selected area diffraction pattern taken from the Ge₃₃Te₆₇ film heated to 235 °C. It is noted that the metastable GeTe₂ film has large grains with (111) preferred orientation, which corresponds to the XRD result. The high electrical resistance of the GeTe₂ film may be due to its crystal structure which is

145

supposed to be isomorphic to that of β-cristobalite SiO$_2$ [8]. However, the metastable GeTe$_2$ compound starts to decompose into α-GeTe and Te phases accompanying a large electrical resistance drop even by a small increase in temperature as confirmed by the XRD measurement (Fig.2(c)). After the decomposition, the electrical resistance of the two-phase α-GeTe + Te slightly increases with increasing temperature.

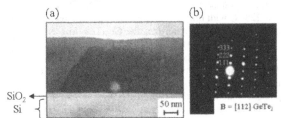

Figure 3. (a) Cross-sectional TEM image and (b) a selected area diffraction pattern taken from the Ge$_{33}$Te$_{67}$ film heated to 235 °C.

B. Phase change process in Ge$_{100-x}$Te$_x$ (x: 46-94) amorphous films

From these results, the compositional dependence of the crystallization temperature and the crystallization process are summarized in Fig. 4. The filled black circles and triangles are the first and the second crystallization temperatures, respectively. They were obtained by finding the minimum value of the first derivative of the electrical resistance with temperature. In the Ge$_{33}$Te$_{67}$ film, the decomposition temperature from the GeTe$_2$ single phase to the GeTe + Te two-phase was adopted as the crystallization temperature, because the crystallization temperature from the amorphous to the GeTe$_2$ cannot be detected clearly and the decomposition occurs just after the crystallization.

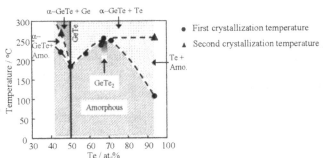

Figure 4. Compositional dependence of the crystallization temperature and processes.

The phase change process of the amorphous Ge$_{100-x}$Te$_x$ (x: 46-94) are categorized by the five regions as listed in Table 1. The compositional dependence of the first crystallization temperature is in good agreement with that in the previous study obtained by differential scanning calorimetry (DSC) [9]. It seems that the second crystallization temperature increases with the Ge content in the region (1) but hardly changes in the region (5). The two-stage crystallization suggests two-stage memory devices, which has also been reported in $Ge_{15}Te_{85}$ film by Messier and Roy [10].

Table 1. The phase change process of the amorphous $Ge_{100-x}Te_x$ films.

Region	Te Composition range	Phase change process
(1)	$46 \leq x < 50$	$a\text{-}Ge_{100-x}Te_x \rightarrow \alpha\text{-}GeTe + amorphous \rightarrow \alpha\text{-}GeTe + Ge$
(2)	$x = 50$	$a\text{-}Ge_{100-x}Te_x \rightarrow \alpha\text{-}GeTe$
(3)	$50 < x < \sim67$	$a\text{-}Ge_{100-x}Te_x \rightarrow \alpha\text{-}GeTe + Te$
(4)	$x \approx 67$	$a\text{-}Ge_{100-x}Te_x \rightarrow GeTe_2 \rightarrow \alpha\text{-}GeTe + Te$
(5)	$\sim67 < x \leq 94$	$a\text{-}Ge_{100-x}Te_x \rightarrow Te + amorphous \rightarrow Te + \alpha\text{-}GeTe$

"a-" indicates the amorphous phase.

B. Activation energy for the first crystallization

The activation energy of the first crystallization is calculated for $Ge_{54}Te_{46}$, GeTe, $Ge_{33}Te_{67}$ and Ge_6Te_{94} using the Kissinger method given as;

$$\ln\left(\frac{dT/dt}{T_c^2}\right) = -\frac{E_a}{k_bT_c} + C . \tag{1}$$

Here, E_a is the activation energy for crystallization, T_c is the first crystallization temperature at a given heating rate dT/dt, k_b is the Boltzmann constant and C is a constant. The E_a was evaluated by measuring the T_c at various heating rates of 2 to 50 °C/min. It is found that the T_c shifts to higher temperature with increasing the heating rate. Figure 5 shows the calculated E_a as a function of the composition. The compositional dependence of the E_a shows a similar tendency to that of the first crystallization temperature. The E_a increases with decreasing the Te content from the Te-rich portion, which is similar to the result reported by Koban et al. [11]. By further decrease in the Te content, the E_a decreases after reaching the maximum point at around Te $\approx$ 67 at.%. Moreover, in the Ge-rich side, the E_a increases with increasing the Ge content, which may be related to the increment of the melting point. In the $Ge_{100-x}Te_x$ ($x \geq 50$) films with a lower melting point than the GeTe compound, the $Ge_{33}Te_{67}$ film shows the highest E_a value, which means that the amorphous state of this composition is most stable. It is supposed that the high crystallization temperature and high activation energy of the $Ge_{33}Te_{67}$ is caused by the existence of the metastable $GeTe_2$ phase. Therefore, the $Ge_{33}Te_{67}$ film with high crystallization and low melting point of about 580 °C can be expected as a candidate material for PCRAM with a low energy consumption.

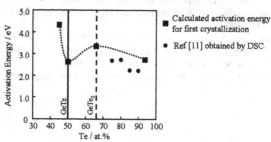

Figure 5. Compositional dependence of the activation energy for first crystallization.

147

CONCLUSIONS

In this research, the relationships of the electrical resistance and the crystal structure change were investigated in wide composition range of the $Ge_{100-x}Te_x$ (x: 46-94). The obtained results were as follows:

(1) The first crystallization temperature increased with decreasing the Te content from the Te-rich portion and then decreased after reaching the maximum point. The $Ge_{33}Te_{67}$ thin film with the metastable $GeTe_2$ phase shows the highest crystallization temperature. In the Ge-rich Ge-Te film, the first crystallization temperature increased with increasing the Ge content.

(2) In the $Ge_{100-x}Te_x$ films with the content range of x < 50 and x > 67, the two-stage drastic changes in the electrical resistance occurred with the two-stage crystallization.

(3) The composition dependence of the activation energy for the first crystallization showed a similar tendency to that of the first crystallization temperature. The $Ge_{33}Te_{67}$ film exhibited the highest crystallization temperature and activation energy.

ACKNOWLEDGMENT

This work was supported by "The Murata Science Foundation" and "Tohoku University Exploratory Research Program for Young Scientists".

REFERENCES

[1] N. Yamada, E. Ohno, K. Nishiuchi, N. Akahira and M. Takao, *J. Appl. Phys.*, **69**, 2849 (1991).

[2] M. H. R. Lankhorst, B. W. S. M. M. Ketelaars and R. A. M. Wolters, *Nat. Mat.*, **4**, 347 (2005).

[3] Y. C. Chen, C. T. Rettner, S. Raoux, G. W. Burr, S. H. Chen, R. M. Shelby, M. Salinga, W. P. Risk, T. D. Happ, G. M. McClelland, M. Breitwisch, A. Schrott, J. B. Philipp, M. H. Lee, R. Cheek, T. Nirschl, M. Lamorey, C. F. Chen, E. Joseph, S. Zaidi, B. Yee, H. L. Lung, R. Bergmann and C. Lam, *IEDM Tech Dig.*, 346910. 1 (2006).

[4] M. Chen, K. A. Rubin and R. W. Barton, *Appl. Phys. Lett.*, **49**, 502 (1986).

[5] L. E. Shelimova, N. Kh. Abrikosov and W. Zhdanova, *Zh. Neorg. Khim.*, **10(5)**, 1200 (1965) [*Russ. J. Inorg. Chem.*, **10(5)**, 650 (1965)].

[6] M. M. Abdel-Aziz, *Appl. Sur. Sci.*, **253**, 2059 (2006).

[7] T. Chattopadhyay, J. X. Boucherle and H. G. von Schnering, *J. Phys. C: Solid State Phys.*, **20**, 1431 (1987).

[8] H. Fukumoto, K. Tsunetomo and T. Imura, *J. Phys. Soc. Jpn.*, **56**, 158(1987).

[9] D. J. Sarrach, J. P. DeNeufville and W. L. Haworth, *J. Non-Cryst. Solids*, **22**, 245 (1976).

[10] R. Messier, R. Roy, *Mat. Res. Bull.*, **6**, 749 (1971).

[11] I. Kaban, E. Dost and W. Hoyer, *J. Alloys and Compounds*, **379**, 1661 (2004).

Mater. Res. Soc. Symp. Proc. Vol. 1160 © 2009 Materials Research Society

Evolution of the Transrotational Structure During Crystallization of Amorphous Ge2Sb2Te5 Thin Films

E. Rimini[1,2], R. De Bastiani[1,3], E. Carria[1,3], M. G. Grimaldi[1,3], G. Nicotra[2], C. Bongiorno[2], C. Spinella[2]

[1]Dipartimento di Fisica ed Astronomia, Università di Catania, 64 via S. Sofia, I-95123 Catania, Italy
[2]CNR-IMM, Stradale Primosole, 50, 95121 Catania, Italy
[3]MATIS CNR-INFM

ABSTRACT

The crystallization of amorphous Ge2Sb2Te5 thin films has been studied by X-ray diffraction (XRD) and transmission electron microscopy (TEM). The analysis has been performed on partially crystallized films, with a surface crystalline fraction (f_S) ranging from 20% to 100%. XRD analysis indicates the presence, in the partially transformed layer, of grains with average lattice parameters higher than that of the equilibrium metastable cubic phase (from 6.06 Å at f_S=20% to 6.01 Å at f_S=100%). The amorphous to crystal transition, as shown by TEM analysis, occurs through the nucleation of face-centered-cubic crystal domains at the film surface. Local dimples appear in the crystallized areas, due to the higher atomic density of the crystal phase compared to the amorphous one. At the initial stage of the transformation, a fast bi-dimensional growth of such crystalline nucleus occurs by the generation of transrotational grains in which the lattice bending gives rise to an average lattice parameter significantly larger than that of the face-centered-cubic phase in good agreement with the XRD data. As the crystallized fraction increases above 80%, dimples and transrotational structures start to disappear and the lattice parameter approaches the bulk value.

INTRODUCTION

Chalcogenide alloys, particularly Ge-Sb-Te (GST), due to the very remarkable difference of the optical reflectivity and electronic conductivity between the amorphous and the crystalline phase, are used as rewriteable layer in Digital Versatile Disc (DVD) and Blu-Ray Disc (BD) and are considered good candidates to replace silicon for the fabrication of ultra-scaled non volatile memory devices. In these materials, the phase–change data storage concept is based on the reversible amorphous to crystal (a-c) transition induced by nanosecond laser pulse or voltage (current) pulse [1] respectively.

The formation of transrotational crystals, during the initial stage of the amorphous to face centered cubic crystal has been reported and attributed to the different density of the amorphous and crystalline phases (~ 6 %) [2-4]. However, their structural evolution, from the early stages of nucleation to the complete crystallization of the GST film, has never been reported before. In this work X-ray diffraction measurements are combined with transmission electron microscopy analyses to investigate in detail the evolution of the crystalline grains during the different steps of the a-c transition.

EXPERIMENTAL

$Ge_2Sb_2Te_5$ amorphous films, 50 nm thick, were prepared by R.F. sputter deposition at room temperature using a stoichiometric target, over a 100 nm SiO_2 layer thermally growth on (001) silicon wafers. The kinetics of the a-c transition, during annealing at 122 ± 0.1 °C, was monitored by *in-situ* time resolved reflectivity (TRR) [5] using a low power (5 mW) He–Ne laser ($\lambda = 633$ nm). The reflectivity increase, during the a-c transformation, has been converted into the crystalline fraction, f_S, by using the effective medium approximation [6]. Since the absorption coefficient of the amorphous and crystalline GST, for at $\lambda = 633$ nm, is about 25 nm^{-1} and 12 nm^{-1}, respectively, the probed thickness never exceeds half of the GST film and f_S can be thought as a *surface crystalline fraction*. The annealing was then stopped at different time intervals to get samples with f_S equal to 20, 40, 60, 80, and 100%, respectively. The structure and morphology of the partially crystallized samples were investigated by X-ray diffraction and transmission electron microscopy. XRD measurements were performed using a SIEMENS D5005 diffractometer (Cu K_α line) in the configuration for thin film analysis, in which the X-ray beam hits the sample surface at a grazing angle (0.5°). Transmission electron microscopy, both in plan view and cross-sectional configuration, was performed using a JEOL JEM 2010F microscope operating at 200 kV and equipped with ultra high-resolution objective lens pole piece. Samples for plan-view images were thinned by backside mechanical lapping, followed by ion milling using a 3 keV Ar^+ beam incident at an angle of 8°. Under these conditions the thinning procedure does not induce appreciable extra-heating of the sample and consequent modifications of his structure. For the cross-sectional analyses we used a focused ion beam (FIB) apparatus after protecting the sample surface by a Pt sputtered layer which prevents any sample damage during the FIB milling.

RESULTS AND DISCUSSION

The XRD patterns of the amorphous and crystallized GST films are shown in Fig. 1(a). The pattern of the amorphous film shows a broad bump centered at $2\theta \sim 30°$. In the crystallized samples, the diffraction peaks due to the lattice planes of the face centered cubic (fcc) crystal appear, and their intensity increases with the fraction of crystalline material. Contemporaneously, the bump associated to the amorphous phase reduces until it disappears in the sample with $f_S =$ 100%. A careful analysis of these data indicates that the angular positions of the (200) and (220) peaks shifts toward larger angles as f_S increases. Such a shift is definitely greater than the error affecting these measurements as can be inferred by looking at the position of the Si peak (at 51.5°), coming from the substrate, that remains constant during the transformation. The dashed lines in Fig. 1(a) identify the angular position of the (200) and (220) peaks, expected for bulk fcc GST ($a = 6.01$ Å) [9], and reached when $f_S = 100\%$. In Fig.1 (b) the measured lattice parameter from the XRD patterns is reported as a function of f_S. It starts to decrease toward the bulk value at the late stage of the transformation. In addition the FWHM of the main diffraction peaks (200) and (220) is quite large, ranging from 0.65° at 20% of surface crystallized fraction to 0.3° at 100% fraction. The last value corresponds, according to the Scherrer formula, to grain size of about 40 nm size.

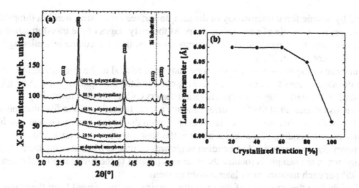

Figure 1. (a) X-ray diffraction of as-deposited amorphous Ge₂Sb₂Te₅ film, partially crystallized and fully polycrystalline cubic phase. The peaks positions related to the diffraction of (200) and (220) planes of Ge₂Sb₂Te₅ structure and the Si substrate contribution are marked with broken vertical lines. (b) Dependence of the lattice constants on the crystallized fraction (cubic phase).

The STEM plan view micrographs of the samples with f_S = 20% (a), 40% (b), 80% (c) and 100% (d), are shown in Fig. 2. The micrographs reveal the presence of GST crystal grains whose density and average size increase with the annealing time until their complete coalescence. At about 100% the average grain size amounts to 300 nm. These grains exhibit the typical contrast of transrotational structures characterized by dark fringes corresponding to bending contours which move laterally by tilting the sample around axes parallel to the fringes themselves [6,7]. Transrotational grains are dominant in the partially crystallized samples [Figs. 2(a,b,c)] and they almost disappear as f_S = 100% [Fig. 2(d)].

Figure 2 STEM plan-view micrographs of as-deposited amorphous Ge₂Sb₂Te₅ films, annealed at 122 ± 0.1 °C with f_S = 20% (a), 40% (b) 80% (c) and 100% (d) respectively. The grains exhibit the typical contrast of transrotational structures characterized by dark fringes corresponding to bending contour. Transrotational grains are dominant in the partially crystallized samples and they almost disappear as f_S = 100%.

Another peculiarity visible in Figs. 2(a,b,c) is the presence of bright and dark small spots (20 nm wide), localized at the center of each transrotational crystal grain or surrounded by the amorphous. They have been identified by the contrast change of TEM images during sample

151

tilting and by atomic force microscopy of the sample surface (not shown here) as dimpled crystalline grains, a couple of nanometers deep. At the early stages of the transformation [Fig. 2(a)], they are mostly surrounded by the amorphous whilst they are completely missing in fully crystallized samples [Fig. 2(d)].

The occurrence of the transrotational structures has been related to the relevant difference between the atomic density of the crystal (6.27 g/cm^3) and the amorphous phase (5.87 g/cm^3) [10]. Indeed, at the early stages of crystallization the crystalline clusters formed by heterogeneous nucleation at the free surface induce a local reduction of the film thickness (dimpled crystallites), because of the elastic deformation due to the density change. As these crystallites grow a bending of the lattice plane at the lateral a-c interface occurs to minimize the elastic energy due to the strain. The bending angle was estimated by measuring the shift of the dark fringe when the sample is rotated about an axis parallel to the fringe itself and it resulted about 0.05° per each nanometer of lateral displacement.

Further insight into the evolution of the crystalline grains can be inferred from the cross-sectional images shown in Fig. 3 which refer to the samples with f_S = 80% [Fig. 3(a)] and f_S= 100% [Fig. 3(b)], respectively. The Pt layer on top of GST has been deposited for the sample preparation for cross section analysis. In the partially transformed sample [Fig. 3(a)] a double layer structure is evident: the upper layer is crystalline while the bottom one is still amorphous. As the crystallization further proceeds, the crystal grains grow consuming the buried amorphous [Fig. 3(b)] and, at the end of the transformation, the thickness of the completely crystallized film reduces by 4% with respect to its initial value.

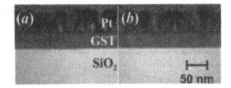

Figure 3. TEM cross-sectional images which refer to the samples with f_S = 80% (a) and f_S= 100% (b), respectively. The Pt layer on top of GST has been deposited to protect the specimen from FIB sample preparation. (a) A double layer structure is evident: the upper layer is crystalline while the bottom one is still amorphous. (b) As the crystallization further proceeds, the crystal grains grow consuming the buried amorphous and, at the end of the transformation, the thickness of the completely crystallized film reduces by 4% with respect to its initial value.

Taking into account the measured angular bending, the size and the estimated thickness of the grains, one can evaluate the average lattice parameter, given by the semi-sum of the distorted lattice parameter on the top of the grain and the bulk lattice parameter on the bottom of the grain respectively. In partially crystallized samples, for grains of about 25 nm lateral size, the average lattice parameter amounts to 6.07 Å in good agreement with the XRD data. Moreover due to the bending, only a limited amounts, about 10%, of crystalline volume contributes to the diffraction thus reducing the size of the diffracting crystal and increasing the FWHM of the diffraction peaks.

From these experimental observations we can conclude that in our explored temperature regime the crystallization of amorphous GST films occurs by the nucleation of small fcc crystal clusters on the film surface which induces a local depression of the surface itself due to the higher atomic density of the crystal phase [Fig. 4(a)].

Figure 4. Schematic representation of the crystal nucleation of amorphous GST films. (a) Nucleation occurs by the nucleation of small crystal on the film inducing a local depression of the surface itself due to the higher atomic density of the crystal phase. (b) The growth of these initial nuclei proceeds by a progressive bending of the lattice planes to partly compensate the increase of the atomic density, thus producing the characteristic transrotational structure which essentially grow parallel to the film surface. (c) As full coalescence is reached the crystallization proceeds in the vertical direction causing the collapse of the transrotational structure.

The growth of these initial nuclei proceeds by a progressive bending of the lattice planes to partly compensate the increase of the atomic density, thus producing the characteristic transrotational structure [Fig. 4(b)] which essentially grow parallel to the film surface. As the coalescence among the transrotational domains complete the crystallization proceeds in the vertical direction causing the collapse of the transrotational structure. At the end of the transformation the whole film is crystallized, the atomic density is the one characteristic of the crystal phase, and consequently the total film thickness reduces [Fig. 4(c)].

CONCLUSIONS

The crystallization of amorphous $Ge_2Sb_2Te_5$ thin films has been studied by transmission electron microscopy and X-ray diffraction. The average lattice parameters of partially transformed samples exceed by 10% that of the equilibrium metastable fcc phase detected at the end of the transformation. The reason for that is the formation, during the crystallization, of transrotational grains in which the lattice bending gives rise to an increase of the average lattice parameter. As the crystallized fraction increases above 80%, the transrotational structures start to disappear and the lattice parameter approaches the bulk value.

ACKNOWLEDGMENTS

The authors thank S. Pannitteri for the useful suggestion on specimen preparation; A. La Mantia and M. Torrisi (STMicroelecronics) for the FIB preparation; A. Pirovano, A. Radualli and A. Gotti from (Numonyx) for useful discussion and for providing us the samples; F. Giannazzo for AFM analysis. Work supported by FIRB Project – RBIP06Y5JJ

REFERENCES

[1] M.Wuttig and N. Yamada, Nature Mater. 6, 824 (2007)
[2] B. J. Kooi and J. Th. M. De Hosson, J. Appl. Phys. 92, 3584 (2002).
[3] B. J. Kooi and J. Th. M. De Hosson, J. Appl. Phys. 95, 4714 (2004).
[4] J. Kalb, C.Y. Wen, F. Spaepen, H. Dieker and M. Wuttig, J. Appl. Phys. 98, 054902 (2005).

[5] R. De Bastiani, A. M. Piro, M. G. Grimaldi, E. Rimini, G. A. Baratta, and G. Strazzulla, Appl. Phys. Lett. 92, 241925 (2008)

[6] D. Kim, F. Merget, M. Laurenzis, P. H. Bolivar, and H. Kurz, J. Appl. Phys. 97, 083538 (2005)

[7] N. Yamada, T. Matsunaga, J. Appl. Phys. 88, 7020 (2000)

[8] A. Alberti, C. Bongiorno, B. Cafra, G. Mannino, E. Rimini, T. Metzger, C. Mocuta, T. Kammler, and T. Feudel, Acta Crystallogr., Sect. B: Struct. Sci. B61, 486 2005

[9] V. Yu Kolosov, A. R. Tholén Acta Mater. 48, 1829 (2000)

[10] M. Wuttig, R. Detemple, I. Friedrich, W. Njoroge, I. Thomas, V. Weidenhof, H.-W. Woltgens, S. Ziegler, J. Magn. Magn. Mater. 249, 492 (2002).

Phase Change RAM II

Mater. Res. Soc. Symp. Proc. Vol. 1160 © 2009 Materials Research Society   1160-H13-02

Field Induced Crystal Nucleation in Chalcogenide Phase Change Memory

Marco Nardone[1], Victor G. Karpov[1], Mukut Mitra[2], ILya V. Karpov[3]
[1] University of Toledo, Dept. of Physics and Astronomy, Toledo, OH 43606
[2] University of Pennsylvania, Materials Science and Engineering, Philadelphia, PA 19104
[3] Intel Corporation, RN3-01, 2200 Mission College Blvd., Santa Clara, CA 95052

ABSTRACT

A summary is presented of our theoretical and experimental work over more than two years related to switching in chalcogenide glass phase change memory. As a significant addition to the well known experiments, we have studied switching under considerably lower voltages and elevated temperatures, as well as the statistics of switching events and relaxation oscillations. Our analytical theory, based on field induced crystal nucleation, predicts all of our observed features and their dependencies on material parameters.

INTRODUCTION

Phase change memory (PCM) devices are based on the phenomenon of switching in chalcogenide glasses [1]. Mainstream understanding is that switching is initiated by an electronic 'hot' filament that can (in PCM) or cannot (in threshold switches - TS) trigger crystal nucleation. The implication is that TS remains amorphous in the low resistance state, i.e. a highly conductive glass, while in PCM a crystalline filament forms some time after the hot filament [2].

We have developed an analytical theory of switching in chalcogenide glasses based on crystal nucleation induced by an electric field [3-5]. Our motivations are: a) the statistical nature of the phenomenon; b) the need for a quantitative theory relating device characteristics to material parameters; and c) the understanding of glasses that has accumulated since the 1970's. In conjunction with the theory we have conducted unique experiments to study switching under conditions far beyond the standard. We have also studied the statistics of switching events and relaxation oscillations in PCM devices [6,7]. Here we present a summary of our theoretical and experimental work over the past two years related to field induced nucleation switching.

THE FIELD INDUCED NUCLEATION MODEL

In classical nucleation theory, the formation of a crystal nucleus is governed by the free energy of the system which creates an energy barrier that must be overcome for nucleation to occur. The height of the barrier determines the probability of and, therefore, the induction time required for, the formation of a stable, spherical, crystal nucleus. The induction time is given by

$$\tau = \tau_0 \exp(W / kT) \tag{1}$$

where W is the barrier height, T is temperature, and k is the Boltzmann constant. The pre-exponential term remains ill-defined but is estimated as the characteristic vibration time $\tau_0 \sim 10^{-13}$ s [5]. The classical barrier height and critical radius are $W_0 = 16\pi\sigma^3/(3\mu^2)$ and $R_0 = 2\sigma/\mu$, where σ

and μ are the surface tension and the chemical potential difference between the two phases, respectively.

The field induced nucleation model extends the classical theory in three important ways. First, it considers the effects of strong electric fields ($\sim 10^5$ to 10^6 V/cm) and the polarization of the conductive (crystal) particle. Second, it allows for two degrees of freedom in nucleation such that the embryo can evolve in either the radial direction or the height. Finally, the model considers the disordered structure of glasses which provides a logical mechanism for the observed statistical nature of switching. The starting point for the theory is the free energy of the system given by

$$F = A\sigma \pm \Omega\mu + W_E \qquad (2)$$

where A and Ω are the nucleus area and volume, respectively. The positive sign in front of μ refers to TS, while a negative sign refers to PCM [7]. The difference in sign is related to the chemical stability of the amorphous versus crystalline phase.

The electrostatic energy term [8] is given by $W_E = -\Omega E_0^2 \varepsilon/(8\pi n)$ where E_0 is the uniform field far from the nucleus and ε is the dielectric permittivity of the amorphous host. The depolarizing factor n accounts for the distortion of the electric field due to polarization of the conducting particle. For a spherical particle $n = 1/3$ while for a prolate spheroid of length h and radius R ("needle") it can be approximated as a cylinder with $h \gg R$ giving $n \approx (R/h)^2$ [6]. A more detailed analysis that includes n, A and Ω for a prolate spheroid can be found in Ref. [4].

By inserting A, Ω, and n for a cylinder into Eq. (1), the free energy takes the form $F = 2\pi R h \sigma \pm \pi R^2 h \mu - h^3 E_0 \varepsilon/8$. Because the parameters space here is two-dimensional (R and h), the system can find nucleation pathways through lower barriers than the energy maximum, as shown in figure 1(a). Physically, the filament radius has a lower bound which is accounted for by introducing a minimum radius αR_0, below which the conductive crystalline cylinder does not exist; α is a phenomenological parameter [5]. Given the minimum radius, the embryo can evolve in a multitude of trajectories in the parameter space (R,h), as shown in figure 1(b).

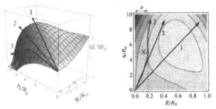

Figure 1. Left: free energy as a function of spheroid radius R and length h. Right: contour plot of free energy. The arrows indicate possible nucleation trajectories and the X is the maximum point (barrier) along the line of minimum radius αR_0.

Analysis of the 2D free energy yields the barrier height for needle-shaped nucleation $W(E) = 2W_0 \alpha^{3/2} E_0/E$; inversely proportional to the field [4]. It was shown that the barrier for nucleation of needle-shaped particles can be significantly lower than that of spherical particles W_0 ($W \sim 0.3 W_0$) [4]. Using $W(E)$ and relating the induction time of the nucleus from Eq. (1) to the switching delay time gives the threshold field E_{th} and delay time τ

$$E_{th} = \frac{1}{\ln(\tau/\tau_0)} \frac{W_0}{kT} \sqrt{\frac{3\pi^3 \alpha^3 W_0}{32 \varepsilon R_0^3}} \quad \text{and} \quad \tau = \tau_0 \exp\left(\frac{W_0}{kT} \frac{V^*}{V}\right) \quad \text{when } V > V^* \approx 0.1\,V \quad (3)$$

where V^* is a minimum voltage below which nucleation switching fails giving up to nucleation of spherical particles [4]. The latter type of nucleation would display a gradual decrease in resistance instead of the abrupt transition of standard PCM switching.

Consideration of electric field screening effects leads to the observed peculiar thickness (l) dependence of the threshold voltage V_{th}. $V_{th}(l)$ is estimated as the product of E_{th} and the length through which it extends, given by the minimum of the amorphous layer thickness l and the bulk electrostatic screening length $l_s = [V e/(2\pi n e)]^{1/2} \sim 1\,\mu m$, where V is the applied voltage and e is the electron charge [9]. Therefore, for thin devices ($l < l_s$) $V_{th} = l E_{th}$ while for thick devices ($l > l_s$) $V_{th} = l_s E_{th}$ [7]; these relations correctly predict the observed values of V_{th} for not only thin PCM [3,4] but also for thin [1,10,11] and thick [10] TS.

Once formed, the nucleus acts as a lightning rod, concentrating the field at its tip and quickly shunting through the amorphous host. However, we showed that an embryo would be unstable against field removal unless it grew its radius beyond $R = 3\pi R_0/16$ [4]. This type of filament instability leads to metastable low resistance states and oscillations in PCM which we observed (see below).

In glasses, the inherent disorder results in nucleation barriers that vary between different microscopic regions in a more or less Gaussian manner [12]. Relating the delay time to the barriers through Eq. (1) results in the log-normal statistical distribution of the delay times and Gaussian distribution of threshold voltages [5]

$$g(\tau) = g_0 \exp\left[-\left(\gamma \ln(\tau/\bar{\tau})\right)^2\right] \quad \text{and} \quad \rho(V_{th}) \propto \exp\left[-\left(V_{th} - \langle V_{th}\rangle\right)^2 / \left(2\Delta V_{th}^2\right)\right] \quad (4)$$

where g_0, g, and $\bar{\tau}$ are functions of barrier height and temperature (cf. Ref. [5]). The terms in angular brackets represent averages and the terms preceded by Δ are dispersions, all of which are field-dependent parameters.

EXPERIMENTAL RESULTS

Our typical experimental setup is illustrated in figure 2. Switching was detected when the voltage V_D across the device dropped, corresponding to branch 3 in the PCM IV curve of figure 2. Data was collected under applied voltages V_A of various magnitudes and pulse times. The current through the PCM device in the set state was controlled by adjusting the load resistance R_L. Here we present a summary of our main results; experimental details are available in the references.

159

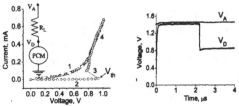

Figure 2. Left: experimental set-up and typical *IV* curve. Right: observed switching event.

We conducted experiments at elevated temperatures (below onset of crystallization) and at applied voltages well below the standard V_{th} for our PCM devices [6]. Long voltage pulse trains were used to facilitate detection of extremely long delay times (compared to typical 10 ns). As shown in figure 3(a), higher voltages and temperatures resulted in shorter delay times. The data also demonstrates 'under-threshold' switching; that is, switching at applied voltages well below the standard V_{th} [6]. Most of the curves in figure 3(a) exhibit an abrupt resistance drop with the exception of the lowest voltage (0.2 V) where a more gradual decline is observed. The latter resembles the behavior of devices under zero bias attributed to the percolation of spherical clusters. The voltage of 0.2 V is close to our estimate of the minimum voltage V^* in Eq. (3). Quantitatively, log τ was found to be approximately linear in $1/T$ and $1/V_D$ as shown in figures 3(b) and (c); consistent with the prediction of Eq. (3).

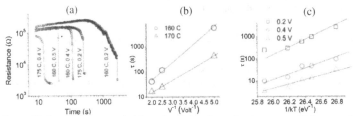

Figure 3. (a) High temperature switching after delay times that increase with lower voltages and lower temperature. Delay time as a function of (b) voltage and (c) temperature.

We also investigated the dependence of V_{th} on τ, T, and device thickness l [3,4]. The results, shown in figure 4, verify our predictions related to Eq. (3).

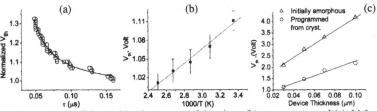

Figure 4. Dependence of threshold voltage on (a) delay time, (b) temperature and (c) thickness.

160

Values of τ that were measured more than 400 times for the same device displayed a broad dispersion which narrowed near the "standard" V_{th}, as shown in figure 5(a). For the same device, figures 5(b) and (c) show that the statistics of the V_{th} and τ were found to be normal and log-normal, respectively, in accordance with Eq. (4).

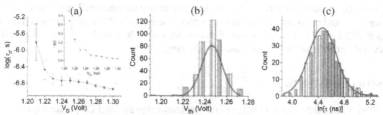

Figure 5. (a) Delay time versus voltage showing the standard deviation. Statistics of (b) threshold voltage and (c) delay time with normal and log-normal distributions, respectively.

Our prediction of the short lived low resistance state (i.e. switching without memory) was verified by using voltage pulses of different widths but equal magnitudes to induce switching [4]. As shown in figure 6(a), a stable resistance change takes place above a certain pulse width proving that time under bias is just as important as time at elevated temperature. Unstable switching behavior is evident when PCM devices exhibit relaxation oscillations. In this domain, PCM behaves like TS in that a minimum holding field E_h is required to maintain the set state (stable filament). Calculation of E_h and consideration of the appropriate screening effects yields the holding voltage [7].

$$V_h = \sqrt{\frac{12\alpha W_0}{\varepsilon R_0}} \text{ for } l < l_s \text{ and } V_h = \frac{l}{l_s}\sqrt{\frac{12\alpha W_0}{\varepsilon R_0}} \text{ for } l > l_s \qquad (5)$$

The thickness dependencies of Eq. (5) are in accordance with previous results [1,13,14] and our own. Figure 6(b) shows a portion of our typical oscillation data demonstrating V_{th} (max. voltage) approximately linear in l and V_h (min. voltage) independent of l, in accordance with the predictions of Eqs. (3) and (5). Furthermore, using typical numerical parameters Eqs. (3) and (5) correctly predict $V_{th} \sim 0.2(l/R_0)V_h$ and $V_h \sim 1$ V, consistent with our observed $V_h = 0.7$ V and historical measurements [13] .

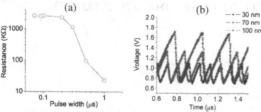

Figure 6. (a) Stable resistance change as a function of voltage pulse width. (b) Relaxation oscillations for different device thicknesses.

CONCLUSIONS

Over the past two years, the field induced nucleation model and our related experiments have lead to a better understanding of many complex PCM switching phenomena, including: 1) under-threshold switching in PCM (also, a minimum voltage below which switching becomes more gradual); 2) the dependence of V_{th}, V_h, and τ on device parameters; 3) the statistical behavior of V_{th} and τ; 4) unstable switching (without memory) in PCM; and 5) the common aspects of PCM and TS of arbitrary size. Future work will include analysis of crystal growth kinetics and further investigation of unstable switching.

ACKNOWLEDGMENTS

Two of us (M.N. and V.G.K.) gratefully acknowledge the Intel Corporation grant.

REFERENCES

1. S. R. Ovshinsky, Phys. Rev. Lett. **21** (20), 1450 (1968).
2. D. Adler, H. K. Henisch, and N. Mott, Rev. Mod. Phys. **50** (2), 209 (1978).
3. V. G. Karpov, Y. A. Kryukov, S. D. Savransky, and I. V. Karpov, Appl. Phys. Lett. **90** (12), 123504 (2007).
4. V. G. Karpov, Y. A. Kryukov, I. V. Karpov, and M. Mitra, Phys. Rev. B. **78** (5), 052201 (2008).
5. V. G. Karpov, Y. A. Kryukov, M. Mitra, and I. V. Karpov, J. Appl. Phys. **104** (5), 054507 (2008).
6. I. V. Karpov, M. Mitra, D. Kau, G. Spadini, Y. A. Kryukov, and V. G. Karpov, Appl. Phys. Lett. **92** (17), 173501 (2008).
7. M. Nardone, V. G. Karpov, D. C. S. Jackson, and I. V. Karpov, Appl. Phys. Lett. **94**, 103509 (2009).
8. L. D. Landau and I. M. Lifshits, *Electrodynamics of Continuous Media.* (Pergamon, New York, 1984).
9. S. M. Sze, *Physics of Semiconductor Devices.* (Wiley, New York, 1981).
10. B. H. Kolomiets, E. A. Lebedev, and I. A. Taksami, Soviet Phys. Semicond. **3**, 267 (1969).
11. P. J. Walsh, R. Vogel, and E. J. Evans, Phys. Rev. **178** (3), 1274 (1969).
12. V. G. Karpov and D. Oxtoby, Phys. Rev. B **54**, 9734 (1996).
13. A. E. Owen and J. M. Robertson, IEEE Trans. Electron Devices **20** (2), 105 (1973).
14. A. C. Warren, IEEE Trans. Electron Devices **20** (2), 123 (1973).

Mater. Res. Soc. Symp. Proc. Vol. 1160 © 2009 Materials Research Society   1160-H13-08

The Influence of Nitrogen Doping on the Chemical and Local Bonding Environment of Amorphous and Crystalline Ge₂Sb₂Te₅

J. S. Washington[1], E. Joseph[2,1], M. A. Paesler[1], G. Lucovsky[1], J. L. Jordan-Sweet[2], S. Raoux[3,2], C. F. Chen[2], A. Pyzyna[2], R. K. Dasaka[2], A. Schrott[2], C. Lam[2], B. Ravel[4] and J. Woicik[4]

[1] Physics Department, North Carolina State University, Campus Box 8202, Raleigh, NC 27695
[2] IBM/Macronix PCRAM Joint Project, T. J. Watson Research Center, Yorktown Heights, NY 10598
[3] IBM Almaden Research Center, San Jose, California 95120
[4] National Institute of Standards and Technology, Gaithersburg, MD

ABSTRACT

Recent interest in phase change materials (PCMs) for non-volatile memory applications has been fueled by the promise of scalability beyond the limit of conventional DRAM and NAND flash memory [1]. However, for such solid state device applications, Ge₂Sb₂Te₅ (GST), GeSb, and other chalcogenide PCMs require doping. Doping favorably modifies crystallization speed, crystallization temperature, and thermal stability but the chemical role of the dopant is not yet fully understood. In this work, X-ray Absorption Fine Spectroscopy (XAFS) is used to examine the chemical and structural role of nitrogen doping (N-) in as-deposited and crystalline GST thin films. The study focuses on the chemical and local bonding environment around each of the elements in the sample, in pre and post-anneal states, and at various doping concentrations. We conclude that the nitrogen dopant forms stable Ge-N bonds as deposited, which is distinct from GST bonds, and remain at the grain boundary of the crystallites such that the annealed film is comprised of crystallites with a dopant rich grain boundary.

INTRODUCTION

As interest in non-volatile memory cells based on phase change materials has increased, research has turned to the optimization of device performance through the control of material properties. Doping allows for the tuning of the properties of Ge₂Sb₂Te₅ (GST), GeSb, and other currently widely accepted PCMs. Nitrogen doping in these materials improves thermal stability [2] and affects crystallite size and transition temperature [3]. Furthermore, finer control of crystallization temperature, T_x, through doping leads to better management of set and reset currents. T_x must be sufficiently above device operation temperature for thermal stability but well below the melting point [4]. The present study employs XAFS to clarify the structural role of nitrogen in N-GST in amorphous and annealed thin films.

EXPERIMENT

One micron thick films of nitrogen doped GST were deposited at room temperature on bare Si wafers and/or glass substrates via RF sputtering from a Ge₂Sb₂Te₅ compound target in a mixture of argon and nitrogen feedgas chemistries. For crystallization, samples were annealed in a helium ambient using a tube furnace; nitrogen containing samples (2.5, 5.0, and 6.1 atomic percent (at %)) were annealed for 30 minutes at 300 °C while the undoped (0.0 at %) sample was

163

annealed at 200 °C for 10 minutes. XAFS measurements were taken at the Advanced Photon Source, on MRCAT Line ID-10 as well as at the National Synchrotron Light Source, on beamline X23A2 in both cases using Si(111) double-crystal monochromators. Germanium (Ge) K edge measurements were taken in florescence, while Antimony (Sb) and Tellurium (Te) measurements were done in transmission. Normalization and background removal of the data were performed using the Athena software package [5].

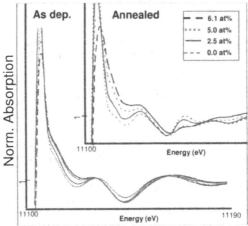

Figure 1: Ge near edge spectra for as deposited and annealed (inset) N-GST samples. N content was determined by x-ray photoelectron spectroscopy (XPS) depth profiling. XPS: monochromatic Al Kα source with a 200 m spot size and charge neutralization. Depth profiling: 1.0 keV Ar+ ions with 3 x 3 mm raster.

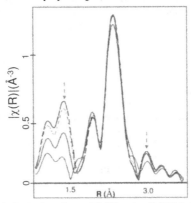

Figure 2: Fourier Transformed EXAFS as dep. N-GST Ge K edge spectra.

The resulting k^2-weighted Fourier transformed EXAFS spectra for all the as-deposited samples were fit using the Artemis analysis packages [5], which is based on the FEFF [6] and IFEFFIT [7] codes.

164

RESULTS AND DISCUSSION

N-GST near edge spectra vs nitrogen flow rate (Figure 1) shows a feature just after the absorption edge which clearly scales as a function of nitrogen atomic concentration, and is correlated to Ge-N bonding in the system. This feature was observed previously, however we cannot confirm that the feature is related to p-p hybridization [8]. Of numerous published works characterizing Germanium nitride, we obtained the near edge spectra for beta-Ge_3N_4, which exhibits an almost matching feature just after the edge [9], shown in figure 3a with an associated EXAFS spectrum shown in figure 3b [10].

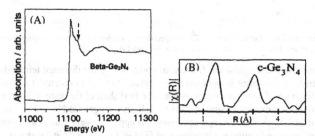

Figure 3: (A) Near Edge Spectra of beta-Ge_3N_4 (arrow indicates the main identifying feature for these) and (B) Fourier transformed EXAFS spectra for beta-Ge_3N_4 (Republished from [10] © AIP 1998).

The authors of that work showed clearly that of the possible bond geometries, the two main peaks in stable Ge_3N_4 are due only to Ge-N bonds and Ge-N-Ge next nearest neighbor interactions, as per a well classified distorted tetrahedral germanium nitride model [11].

Fourier transformed EXAFS Ge K edge spectra for the as-deposited samples (Figure 2) have matching features at similar radial distances to the stable Ge_3N_4 spectra. For completeness, we show the typically ignored surface oxide in the 0.0 at% sample and any resemblance to the Ge-N spectra in the doped samples is coincidental. Also, notice that the Ge-N bond lengths appear at a non physical distance of 1.5 Å due to inherent phase terms that must be solved in the EXAFS equation. A more significant observation from the data above and previous experience with 0.0 at% as deposited samples [12], is that the local structure of amorphous GST bonds appeared unperturbed by nitrogen addition.

This evidence indicates a model of a-GST + nanoscale ordered Ge-N bonds. To test this model, we input into IFEFFIT the distorted tetrahedral unit cell parameters from stable Ge_3N_4 [11] plus possible bond pairs from a-GST (tetrahedrally bonded Ge-Ge, Ge-Te, and Ge-Sb which follow the 8-N rule). Apart from bond distance (R) and coordination number (N), we also found the fraction of bonds in all environments (germanium nitride, surface germanium oxide, or bulk GST). The preliminary results of the fits are shown in Figure 4, where in each case our best fits included Ge-N-Ge contributions - the typical measurement uncertainty in R is 0.00093. Note that all fits shown are on common scales.

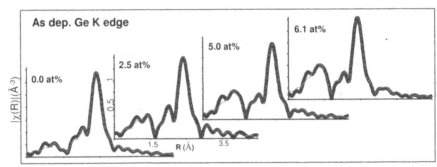

Figure 4: Fourier transformed amorphous Ge K edge spectra of amorphous samples of varying N content as shown. The data (thick dark lines) substantially obscure the model (light gray lines).

The data in table 1 distinguish the fraction of bonds within the distorted tetrahedral germanium nitride environment as opposed to amorphous GST. As expected, this fraction increases as a function of nitrogen concentration. The first shell of the distorted tetrahedral environment contains 2 nitrogen atoms at 1.80 Å, a slightly longer Ge-N bond at 1.82 Å, and a long Ge-N bond at 1.89 Å. The average bond length is approximately 1.83 Å as reported [10]. Also shown are the co-ordination environments of Ge-Ge, Ge-Sb, and Ge-Te as a function of composition. Sb and Te K edge data are not shown, however excellent fits to these edges do not include a co-ordination environment at low R. At these edges, we see only Sb-Ge / Sb-Te bonds from the Sb K edge, and Te-Ge / Te-Sb at the Te K edge (as one expects from a-GST).

Bond	Bond Env. Fraction 2.5%	5.0%	6.1%	N	$\sigma^2(\text{Å}^2)$	Energy shift	R (Å)
Germanium Nitride Ge- N	32%	44%	48%	2.00	0.004		1.80
Ge- N				1.00	0.004		1.82
Ge- N				1.00	0.004		1.89
Ge- N- Ge				2.00	0.011		3.08
Ge- N- Ge				2.00	0.011		3.18
Ge- N- Ge				4.00	0.011		3.24
GST Ge- Ge	68%	56%	52%	0.41 2.5 at% / 0.36 5.0 at% / 0.46 6.1 at%	0.003	4.05 eV- 2.5 at% / 4.02 eV - 5.0 at% / 4.15 eV - 6.1 at%	2.43
Ge- Sb				1.00 2.5 at% / 0.79 5.0 at% / 0.69 6.1 at%	0.003		2.71
Ge- Te				2.59 2.5 at% / 2.86 5.0 at% / 2.85 6.1 at%	0.003		2.60

Table 1: Fit Statistics for the nitrogen containing samples. For germanium nitride, separate rows are indicated for three distinct bonds lengths in the distorted tetrahedra. For GST, each row (Ge-Ge, Ge-Sb, and Ge-Te) indicates one bond length for the three samples of varying doping levels.

All data sets were fit concurrently assuming that R and mean square displacement (σ^2) were identical in all samples, while energy shift (ΔE), passive electron reduction factor (S_o^2), bond environment fraction were sample specific. In this scheme, approximate errors in N and ΔE were typically 10-20% while errors in R and σ^2 were 1-5%. Errors in N denote the range of observed error bars for S_o^2 and the bond environment fraction as calculated by FEFF. The observed errors

166

for Ge-Sb fit variables were typically 5-10% higher in all data sets, indicative of the difficulty in resolving bond statistics about Ge-Sb and Ge-Te nearest neighbors.

Amorphous Ge-N bonds are thermally stable up to 500 °C, while beta-Ge$_3$N$_4$ is stable past 700 °C [13]. In either case, the Ge-N environment is expected to persist post anneal. In figure 5, the Ge-N bonding features observed in the amorphous samples as indicated by the arrows (see figure 2) are retained. Also, the near edge spectra of these samples (inset of figure 1) show a feature just past the edge which scales with nitrogen content.

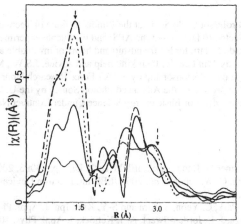

Figure 5: Fourier Transformed EXAFS of annealed N-GST Ge K edge spectra.

This observation leads to a potential supposition for the crystallization behavior of doped GST samples in that GST crystallites would seemingly form within a stable Ge-N environment as the nitrogen doping concentration increases. As confirmation, figure 6 shows TEM images of similar N-GST thin films found in literature and reveals a similar trend, a high Z material (GST) within a low Z matrix (Germanium Nitride) exhibiting a reduction in crystallite size as a function of nitrogen concentration. [14].

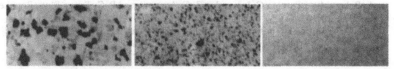

Figure 6: In-Situ TEM of N-GST thin films anneal by hot plate. From left to right – 0.0 at%, 2.9 at%, & 6.8 at%. (Republished from [14] © JJAP 2000)

CONCLUSION

Using a combination of XANES and EXAFS, it was found in this work that nitrogen dopants in N:GST films bond solely to germanium, exhibiting ordered local structure reminiscent of stable Ge$_3$N$_4$. Additionally, the fraction of Ge bonds in this ordered Ge-N

167

environment varied from 32% to 48% as the N concentration within the N:GST films increased from 2.5% to 6.1%. As a function of crystallization, we observed matching near edge as well as EXAFS spectra post anneal indicating that the stable germanium nitride remains separate from GST before and after the anneal (ie: on the grain boundary). In the near future, we will present our detailed results of the annealed samples, as well as an EXAFS study from the nitrogen K edge.

ACKNOWLEDGEMENTS

We are grateful to Professor Carlo Segre at the Illinois Institute of Technology for allowing us to use the MRCAT Sector 10 facilities at the APS, and invaluable experimental assistance. Special thanks to Leonardo Miotti, for his friendship and hours of invaluable assistance and Grant Bunker as well as Jay-Min Lee, for their kind help and advice. J.S.W., M.P., and G.L. additionally acknowledge partial sponsorship by the Air Force Research laboratory under grant no. F29601-03-01-0229 as well as by the Advanced Photon Source by the U. S. Department of Energy, Office of Science, Office of Basic energy Sciences, under Contract No. W-31-109-ENG-38.

REFERENCES

1. Y.C. Chen, C.T. Rettner, S. Raoux et al., IEDM Tech. Dig., p. S30P3, 2006.
2. S. J. Ahn, Y. J. Song, C. W. Jeong et al., 2004 IEEE Int. Electron Devices Meeting, San Francisco, CA, Dec. 2004.
3. H Seo, T. Jeong, J. Park, C. Yeon, S. Kim and S-Y. Kim, Jpn. J. Appl. Phys. 39, 745 (2000).
4. S. Raoux, M. Salinga, J. L. Jordan-Sweet, A. J. Kellock, J. Appl. Phys. 101, 044909 (2007).
5. B. Ravel and M. Newville, J. Synchrotron Rad. 12, 537 (2005).
6. J.J. Rehr, J. Mustre de Leon, S.I. Zabinsky, and T.C. Albers, J. Am. Chem. Soc. 113, 5135 (1991).
7. M. Newville, J. Synchrotron Rad. 8, 322 (2001).
8. Y. Kim, M. H. Jang, et al, App. Phys. Lett. 92, 061910 (2008).
9. C. Bull, P. F. McMillan, J. Itié, and A. Polian, Phys. stat. sol. (a) 201, 5 (2004).
10. I. Chambouleyron and A.R. Zanatta J. Appl. Phys., 84, 1, 1998
11. S. N. Ruddlesden and P. Popper, Acta Cryst. 11, 465 (1958).
12. D. A. Baker, M. A. Paesler, G. Lucovsky, S. C. Agarwal, and P. C. Tayor, Phys. Rev. Lett. 96, 255501 (2006).
13. T. Maeda, T. Yasuda, M. Nishizawa et al., Appl. Phys. Lett. 85, 3181 (2004).
14. H. Seo, T. Jeong, J. Park et al., Jpn. J. Appl. Phys. 39, 745 (2000).

Phase Change RAM III

Mater. Res. Soc. Symp. Proc. Vol. 1160 © 2009 Materials Research Society 1160-H14-04

Local Bonding Asymmetries in Ge-As-Se Glasses

E. Mammadov[1], P. C. Taylor[2], D. Baker[2], D. Bobela[2], A. Reyes[3], P. Kuhns[3], S. Mehdiyeva[1]

[1]Institute of Physics, National Academy of Sciences, Baku, AZ 1143, Azerbaijan
[2]Department of Physics, Colorado School of Mines, Golden, CO 80401, USA
[3]National High Magnetic Field Laboratory, Tallahassee, FL 32000, USA

ABSTRACT

High magnetic fields up to 22 T have been applied to determine the local bonding asymmetries in the $Ge_2As_2Se_7$ and $Ge_2As_2Se_5$ glasses by [75]As NMR. The results are analyzed using computer simulations of the line-shape. The asymmetry parameter η of the electric field gradient at arsenic sites in $Ge_2As_2Se_7$ is found to be about 0.2, indicating that the dominant arsenic structural units in the composition are nearly axially symmetric pyramids. In the $Ge_2As_2Se_5$ glass, however, the [75]As NMR spectrum exhibits no well-resolved structure, revealing the existence of highly asymmetric sites. The experimental data are fitted using the previously obtained distribution of quadrupole coupling constants from pulsed [75]As NQR experiments. These NMR simulations assume a broad distribution of the asymmetry parameter. The results are in agreement with the NQR studies in the same compositions.

INTRODUCTION

Magnetic resonance techniques provide very useful tools to probe the local bonding arrangements in amorphous solids. Nuclear Quadrupole Resonance (NQR) and Nuclear Magnetic Resonance (NMR) spectroscopies have been successfully applied for obtaining structural information in glassy materials [1-8]. Due to various important technological applications As-containing glasses are of special interest. [75]As nucleus (100 % abundant) with spin $I = \frac{3}{2}$ posses a quadrupole moment, which can interact with the electrical field gradient (EFG) created at the nuclear site. This field gradient is strongly influenced by the nearest-neighbor atoms. The interaction between the nuclear quadrupole moment and the EFG can be probed by studying the response of the nuclear spin system to excitation by an rf pulse. For the arsenic nucleus a single transition between doubly degenerate nuclear states occurs at a frequency [9]:

$$ \nu_{NQR} = \frac{e^2 qQ}{2h} \sqrt{1 + \frac{1}{3}\eta^2} , \qquad (1) $$

where $e^2 qQ / h = \nu_Q$ is the so-called quadrupole coupling constant and η is the asymmetry parameter of the EFG. In the case of glasses the NQR line is broadened due to a distribution of the transition frequencies, which occurs because the long-range order is different at different sites due mainly to variations in bond angles. Thus the two parameters, ν_Q and η, can uniquely describe the quadrupole interaction and, correspondingly, the local environment of a given atom.

However, since there is only one transition frequency and Eq. (1) does not depend strongly on η, this parameter is not easily determined just from NQR experiments.

When a magnetic field is applied that is strong enough for the Zeeman interaction to prevail over the quadrupole one (perturbation limit), the NMR powder pattern exhibits two divergences at the central transition (m = ½ to - ½) [10] (Fig.1). The smaller peaks appearing below 17 T and above 21T are the satellite transitions, which we will not discuss here.

In general, the separation Δs between the two divergences depends on η [10]. In the case of arsenic nucleus this separation can be expressed analytically (in the perturbation limit) as [11]:

$$\Delta s = \frac{v_Q^2}{v_O} \frac{\left(\eta^2 - 22\eta + 25\right)}{48}, \qquad (2)$$

where v_O is the spectrometer operating frequency. This separation is strongly dependent on η for a given v_Q. Thus, a combination of the NQR and high field NMR measurements can provide quite accurate values of v_Q and η.

In this work we report high field NMR measurements on $Ge_2As_2Se_7$ and $Ge_2As_2Se_5$ glasses. These compositions are at or above the $(GeSe_2)_x(As_2Se_3)_{1-x}$ tie-line (a threshold for chemical ordering) where Ge, As, and Se atoms exhibit four, three, and two coordinations, respectively, and only Ge-Se and As-Se bonds are present. These compositions are important to the understanding both the possibility of increases in the local coordination numbers and the necessity for like-atom bonds.

EXPERIMENTAL DETAILS

Glassy samples were prepared by conventional melt quenching technique in the Crystal Growth Laboratory of the University of Utah. Glasses were powdered with the grain size no less than 100 μm and placed into 10 mm long and 5 mm OD pyrex tubes to fit the NMR coil.

It is characteristic for As-chalcogenide systems to possess large values of v_Q. The quadrupole coupling constant for a p electron on the As nucleus is ~ 400 MHz [1]. Thus one needs to apply very high magnetic fields to ensure the validity of the perturbation limit. Also these materials exhibit a very broad spectrum in the frequency domain, which is beyond the coverage of a typical NMR spectrometer. To circumvent this problem the line shape has been obtained by sweeping the magnetic field at the fixed spectrometer operating frequency.

High field ^{75}As NMR measurements were performed using a 30 T homogeneous Bitter magnet at the National High Magnetic Field Laboratory. The pulsed NMR spectrometer was operated at 137 MHz which corresponds to a Zeeman field of 18.78 T for ^{75}As nucleus. The spin-echo technique was utilized with a 90° -180° pulse sequence, and the typical 90° pulse width was 4 μs. The NMR measurements were carried out at 80 K using standard NMR probe immersed in liquid nitrogen. The NMR spectrum is constructed by taking the integral intensity of the echo at different values of the magnetic field. In order to increase the signal-to-noise ratio a Carl-Purcell-Meiboom-Gill (CPMG) pulse sequence was used with a train of 95 echos. The echos were added up and the resultant echo intensity is used to map the spectrum.

EXPERIMENTAL RESULTS

The high field NMR data for both compositions at applied fields up to 22T are shown in Fig. 2. The figure shows the central components of both spectra to emphasize the two divergences. These two spectra exhibit completely different lineshapes. One can observe two divergences in the central transition for glassy $Ge_2As_2Se_7$, while the spectrum for $Ge_2As_2Se_5$ exhibits no well-resolved features within the experimental error. The divergences on the NMR spectrum for

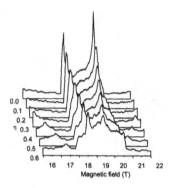

Figure 1. Simulation of the NMR powder patterns for the spin $I = \frac{3}{2}$ without distribution of v_Q at the operating frequency $v_0 = 137$ MHz. The spectrum has been truncated to emphasize the divergences at the central transition. The distance between divergences is determined by η for a constant v_Q.

$Ge_2As_2Se_7$ are separated roughly by 1.5 T. The asymmetry parameter increases for $Ge_2As_2Se_5$ as evidenced by the fact that the divergences get closer to each other [10] (cf., Figs. 1 and 2). NMR powder patterns have been simulated to fit the experimental data. The simulations were performed by averaging the transition frequencies over a large number of random orientations of the applied magnetic field with respect to the principal axes of the EFG. For the distribution of η a Gaussian function of the form $g(\eta) = 0.4 / \Delta \exp[-0.5(\eta - \eta_0 / \Delta)^2]$ has been used, where η_0 corresponds to the peak value of the distribution and Δ is the spread of η values. The resulting pattern was weighted by using the distribution of v_Q obtained from the NQR data (Fig. 3). The NQR spectra also exhibit very broad linewidths, which are typical for amorphous As chalcogenides [8, 12]. These broad spectra provide an accurate distribution of v_Q (Eq. 1). The NQR spectrum for $Ge_2As_2Se_7$ is a single broad line almost symmetrically centered about the 60 MHz. The high frequency tail possibly comes from strains in the glassy network associated with rapid cooling of the glass from the melt or from some bond angles that depart markedly from the average [13]. This asymmetry does not affect the results because the NMR and NQR spectra can be fit self-consistently using the same distributions of ν_Q and η. The center of mass of the NQR spectrum for $Ge_2As_2Se_5$ is shifted to higher frequencies with one intense peak near ~ 71 MHz and another smaller one near ~ 60 MHz.

DISCUSSION

Recent [75]As NQR experiments [14] in the Ge-As-Se glassy system have shown that the local bonding structure in this system radically changes when the composition of the glasses

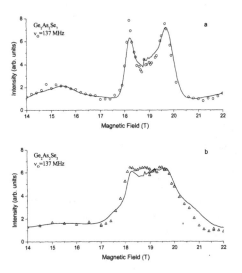

Figure 2. [75]As high field NMR spectra (triangles and circles) for (a) $Ge_2As_2Se_7$ and (b) $Ge_2As_2Se_5$ at the operating frequency, $\nu_0 = 137$ MHz. Data are scaled for the clarity. Solid lines are the calculated powder patterns weighted according to the distributions of ν_Q to fit the experimental data.

contains less Se than that which occurs on the pseudobinary tie-line $(GeSe_2)_x(As_2Se_3)_{1-x}$. The $Ge_2As_2Se_7$ glass is located on this tie-line where only Ge-Se and As-Se bonds occur. On the other hand, the $Ge_2As_2Se_5$ glass is a composition with an excess of Ge and As atoms.

As the NQR lineshape exhibits a broad distribution of transition frequencies the fitted line shown in Fig. 2 was obtained by weighting the calculated data in accordance with the distribution of ν_Q. The value of η determined for $Ge_2As_2Se_7$ is 0.2 (with $\Delta = 0.01$) is very close to that for glassy As_2Se_3 [11]. Indeed, in this ternary system constituent atoms tend to form heteropolar bonds [15] so that in $Ge_2As_2Se_7$ all Ge and As valences are satisfied by Se atoms. Thus As-pyramidal units are homogeneously distributed in the glassy matrix [16]. The breadth of the NQR spectrum of this composition is primarily determined by deviations in the apex bonding angles [1], which contribute to the distribution of EFGs. As the full width at half maximum of the NQR spectrum is about 14 MHz (which is slightly wider than in As$_2$Se$_3$ [8]) the distribution of EFG components is about 22 %. At small values of η ($\leq \frac{1}{3}$) the NQR frequency is independent of η (eq. 1). Thus the distribution of EFG components that is important for the NQR spectrum is the distribution in ν_Q [17]. The distortions in the bond angles of the Ge$_2$As$_2$Se$_7$ are also larger than those in As$_2$Se$_3$, perhaps due to the incorporation of Ge atoms.

In contrast to the Ge$_2$As$_2$Se$_7$ glass the NMR spectrum for glassy Ge$_2$As$_2$Se$_5$ has no well-resolved structure (Fig. 2 (b)). This lineshape cannot be fitted with a small distribution of asymmetry parameter.

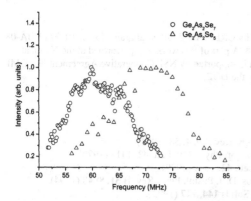

Figure 3. The distribution of ν_Q for ^{75}As nucleus in Ge$_2$As$_2$Se$_7$ (circles) and Ge$_2$As$_2$Se$_5$ (triangles) glasses at 77 K. Data are scaled to allow ease in comparison.

The powder pattern shown in Fig. 2 (b) has been calculated assuming a Gaussian distribution of ν_Q for the two NQR peaks near ~ 71 and ~ 60 MHz and a broad distribution of the asymmetry parameters. Despite imperfections in the fitting it is clear that a significant contribution to the lineshape comes from arsenic sites with high values of η. This result suggests that the local structure of the glass contains highly distorted structural units with a significant portion of arsenic-arsenic homopolar bonds [12]. This latter conclusion comes from the attribution in the As-Se glasses of the NQR peak near 70 MHz to the presence of As-As bonds. The probability of As-Se bonds is drastically decreased when going from Ge$_2$As$_2$Se$_7$ to Ge$_2$As$_2$Se$_5$ [18]. In order to satisfy all their bonding requirements the As atoms form As-As homopolar bonds and the majority of them have two homopolar bonds per site. The contribution to the NMR line from nuclear sites with different values of η is consistent with the existence of complex molecular agglomerations or nodules in this composition.

We have presented the results of ^{75}As high-field NMR measurements that reveal important information on the asymmetries of the local bonds in the $Ge_2As_2Se_7$ and $Ge_2As_2Se_5$ glasses. The results show that all of the As sites in the former composition are nearly axially symmetric and that the arsenic atoms are bonded to selenium atoms in pyramidal structural units. On the contrary, in the $Ge_2As_2Se_5$ most As atoms form highly distorted homopolar bonds, which produce the featureless NMR spectrum. A calculated powder pattern shows that highly asymmetric arsenic sites contribute to the lineshape for this composition. This result suggests that the local bonding structure in glassy $Ge_2As_2Se_5$ is more disordered than that in $Ge_2As_2Se_7$. The results are consistent with those previously obtained in As-Se glassy system.

ACKNOWLEDGEMENTS

This work is partially supported by CRDF/ANSF bilateral grant No. AZP1-3114-BA-08 and by an NSF grant No. DMR-0073004. A part of this work was performed at the National High Magnetic Field Laboratory, which is supported by NSF Cooperative Agreement No. DMR-0084173, by the State of Florida, and by the DOE.

REFERENCES

1. Mark Rubinstein and P. C. Taylor, Phys. Rev B **9**, 4258 (1974).
2. G. E. Jellison, G. L. Petersen, P. C. Taylor, Phys. Rev. Lett. **42**, 1413 (1979).
3. J. Szeftel and H. Alloul, Phys. Rev. Lett. **42**, 1691 (1979).
4. S. R. Rabbani, N. Caticha, J. G. Santos, D. J. Pusiol, Phys. Rev. B **51**, 8848 (1994).
5. D. Mao and P. J. Bray, J. Non-Cryst. Solids **144**, 217 (1992).
6. I. Korneva, M. Ostafin, N. Sinyavsky, B. Nogaj, M. Maćkowiak, Solid State Nucl. Mag. Res. **31**, 119 (2007).
7. S. J. Gravina, Phillip J. Bray, G. L. Petersen, J. Non-Cryst. Solids **123**, 165 (1990).
8. E. Ahn, G. A. Williams. P. C. Taylor, D. G. Georgiev, P. Boolchand, B. E. Schwickert, R L. Cappelletti, J. Non-Cryst. Solids **299-302**, 958 (2002).
9. T. P. Das and E. L. Hahn, *Nuclear Quadrupole Resonance Spectroscopy*, Solid State Physics Suppl. 1, edited by F. Seitz and D. Turnbull (Academic, New York, 1958).
10. P. C. Taylor, J. F. Baugher, H. M. Kriz, Chem. Rev. **75**, 203 (1975).
11. P. C. Taylor, P. Hari, A. Klienhammes, P. L. Kuhns, W. G. Moulton, N. S. Sullivan, J. Non-Cryst. Solids **227-230**, 770 (1998).
12. Z. M. Saleh, G. A. Williams, and P. C. Taylor, Phys. Rev. B **40**, 10557 (1989).
13. E. Mammadov. P. C. Taylor (to be published).
14. E. Mammadov and P. C. Taylor, J. Non-Cryst. Solids **354**, 2732 (2008).
15. C. Rosenhahn, S. Hayes, G. Brunklaus, and H. Eckert in *Phase Transitions and Self-organisation in Electronic and Molecular Networks* ed. by J. C. Phillips and M. F. Thorpe, Springer, New-York (2001) 450 p.
16. R. Zallen, *The Physics of Amorphous Solids*, Wiley-VCH, Weinheim (2004).
17. T. Su, P. Hari, E. Ahn, P. C. Taylor, P. L. Kuhns, W. G. Moulton, and N. S. Sullivan, Phys. Rev. B **67**, 085203 (2003).
18. Z. U. Borisova, *Glassy Semiconductors* (Plenum Press, New York, 1981).

Mater. Res. Soc. Symp. Proc. Vol. 1160 © 2009 Materials Research Society
1160-H14-05

Epitaxial Phase Change Materials: Growth and Switching of Ge2Sb2Te5 on GaSb(001)

Wolfgang Braun[1], Roman Shayduk[1], Timur Flissikowski[1], Holger T. Grahn[1], Henning Riechert[1], Paul Fons[2] and Alex Kolobov[2]
[1]Paul-Drude Institute for Solid State Electronics, Berlin, Germany
[2]Center for Applied Near-Field Optics Research, National Institute of Advanced Industrial Science and Technology, Tsukuba, Japan

ABSTRACT

Epitaxial Ge2Sb2Te5 has been successfully grown on GaSb(001) by molecular beam epitaxy. The films show a tendency for void formation and rough morphology, but at the same time a very strong epitaxial orientation, cubic structure and a sharp interface to the substrate. The layers can be reversibly switched between the crystalline and amorphous phases using short laser pulses.

INTRODUCTION

The continued scaling of non-volatile flash memory will reach fundamental limits in the very near future. Beyond the 32 nm feature size, new memory concepts are required to overcome this barrier. Phase change materials such as GeTe--Sb2Te3 alloys (GST) and in particular Ge2Sb2Te5, are among the most likely candidates for this task [1]. This is due to their good scaling and especially the high switching speed that approaches current DRAM technologies, allowing the vision of a single, nonvolatile memory that replaces both flash and DRAM. Despite the widespread use of GST in optical memories for many years, the basic switching mechanism and the origin of the fast switching are still not well understood. Extended x-ray absorption fine structure (EXAFS) measurements reveal a pronounced change in the nearest neighbor distance and coordination of the Ge atom, which is consistent with an umbrella-flip like site change of Ge [2]. Such a short-range motion could explain the observed high switching speed. On the other hand, the long-range strucural changes observed by diffraction are dramatic. Despite its good crystallinity, the Debye-Waller factors of the metastable cubic phase are very large [3]. The strong variations in the long-range order are a concern for the reliability of GST in memory applications. Another very interesting question is the ultimate scaling limit [4], which determines the future potential of any future memory material.

GST is usually sputtered in polycrystalline or amorphous form without taking care about the crystallinity and lattice constant of the substrate. This complicates the structural analysis, especially of the crystalline phase, since multiple crystallite orientations are present. To be able to study the structural changes during switching in more detail, we have begun to grow GST in an epitaxial orientation on single crystalline substrates [5,6] and investigate the properties of such films.

Here, we report on the growth, film structure and switching of epitaxial GST films on GaSb(001).

EXPERIMENT

The GST films are grown *in situ* in a molecular beam epitaxy (MBE) system at our x-ray diffraction beamline at the synchrotron BESSY II of the Helmholtz Center, Berlin [7]. The pressure during deposition is around 10^{-9} mbar. We synthesize GST from elemental fluxes of high-purity MBE-grade Ge, Sb and Te. Typical source temperatures are 1092, 536 and 346 °C, respectively. The growth rate at unity sticking is around 0.4 nm/min. The on-axis GaSb substrates are prepared in the usual way by desorbing the native oxide at 530 °C and then growing a homoepitaxial buffer layer at substrate temperatures between 300 and 360 °C and a V-III ratio of around 2.5. The growth is monitored *in situ* by reflection high-energy electron diffraction (RHEED) and x-ray diffraction. The substrate temperature is monitored by a thermocouple located near the back of the substrate.

GST has been grown at substrate temperatures ranging from room temperature to 370 °C by simultaneously opening all three shutters on the quenched GaSb(001) surface. The composition of the GST is determined and adjusted by x-ray fluorescence measurements on reference samples deposited on Si at room temperature.

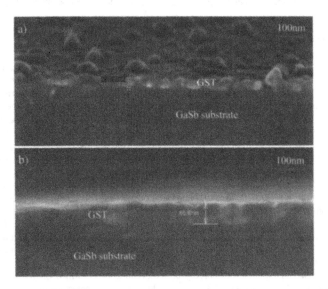

Figure 1. Epitaxial $Ge_2Sb_2Te_5$ films grown on GaSb(001) at (a) 190 °C and (b) 180 °C. The thicknesses of the continuous films are 35 nm in (a) and 65 nm in (b).

DISCUSSION

Scanning electron microscopy (SEM) images of two typical samples are shown in Fig. 1. The sample in Fig. 1a was grown at a substrate temperature of 190 °C for 72 min and the sample in Fig. 1b was grown at 180 °C for 157 min. In both cases, the films have a nonuniform structure, with large crystallites several times the height of the continuous film protruding out of the layer for the film shown in Fig. 1a. Data on the composition and density of several layers are shown in Fig. 2. Due to the stability and reproducibility of the fluxes from the effusion cells, the composition calibration on the reference samples is straightforward (Fig. 2a). Since the GaSb substrate contains Sb, however, a reliable composition determination is not possible for films grown on GaSb. Films grown at substrate temperatures above 130 °C, where the sticking coefficient is less than unity, showed less Te compared to Ge than the nominal 5/2 ratio. Experiments on Sb free substrates are planned to resolve this issue.

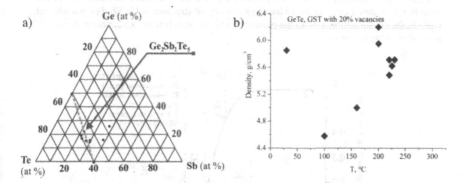

Figure 2. (a) Film composition determined from x-ray fluorescence analysis of amorphous reference samples and (b) densities of films with nominally identical composition determined by x-ray reflectivity as a function of substrate temperature.

The density data shown in Fig. 2b reveal values consistently lower than expected for crystalline cubic $Ge_2Sb_2Te_5$. Whereas the amorphous material deposited at 30 °C is close to the nominal value, the density is significantly lower as soon as crystalline material is formed. This is the case even at growth rates much smaller than the supplied total flux. GST obviously has the tendency to form vacancies in excess of the required 20% of the cubic structure when deposited above the crystallization temperature. The effect seems to be strongest at 100 °C. This may be due to a combination of two effects: the low mobility of adatoms at these temperatures and the tendency to form non-compact structures as soon as cubic or hexagonal structure elements appear. The problem for epitaxy lies in the structure itself: whereas the 20% of vacancies are arranged in layers normal to the c axis in the hexagonal structure, these vacancies are randomly distributed within the Ge-Sb sublattice according to the current structure model for the metastable cubic phase. The vacancies are necessary to stabilize the bulk structure. On the other hand, one can

179

imagine that they are difficult to form at the surface in an epitaxial growth process. The enclosure of a stable vacancy probably requires the simultaneous arrangement of several atoms around it and therefore may requires the coincident arrival of several atoms at the corresponding site. This makes it a low probability process. The growth window to obtain crystalline epitaxial films therefore is quite narrow [6] and the growth rate decreases rapidly at rather low temperatures as shown in Fig. 3. The growth around 200 °C proceeds in an unusual 'incubated epitaxial' mode[5]: Initially, an amorphous film with a thickness corresponding to between one and two cubic lattice constants forms, and then crystallization takes place in that film. After the crystallization step, the film continues to grow in crystalline form since the cubic structure is continuously observed in RHEED without a significant diffuse background due to amorphous material. The curve fitted to the data in Fig. 3 is the difference between the vapor pressure of GST and the supplied flux. The fact that the data points can be fitted by a single curve indicates that GST sublimates congruently at these temperatures and that no phase separation takes place. At substrate temperatures where the growth rate is strongly below the value defined by the incident flux, however, the above considerations would favor the growth of Sb deficient material as cubic GeTe has the same crystal structure as $Ge_2Sb_2Te_5$, but without voids in the Ge sublattice. Due to the presence of Sb in the substrate, we cannot test this possibility at the moment.

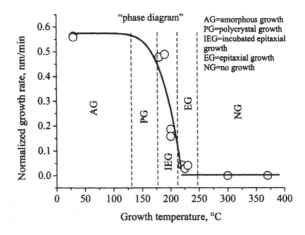

Figure 3. Growth rate vs. substrate temperature for films deposited with a nominal $Ge_2Sb_2Te_5$ composition. Crystalline films are only obtained at temperatures where the sticking coefficient is less than one. The solid line is a fit obtained from the difference between the GST vapor pressure and the deposition flux.

We do not find intermixing or a reaction with the substrate. SEM data such as the ones shown in Fig. 1 always show an abrupt interface. Exposure of the GaSb surface to the GST flux at temperatures above 250 °C leads to a change in surface reconstruction, no growth, but also no roughening of the surface. Instead, the surface tends to become even smoother with large terraces separated by monoatomic steps [6].

As-grown films can even be desorbed and the surface returned to this state. Such an experiment is shown in Fig. 4. The three panels of the figure show the substrate temperature, the chamber pressure close to the substrate and selected RHEED patterns at various stages of the process.

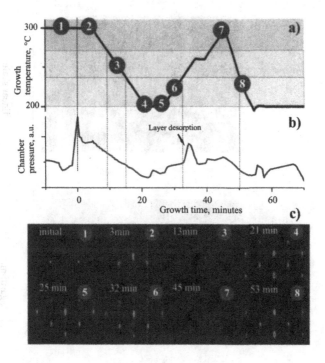

Figure 4. Growth and desorption of GST under combined constant Ge, Sb and Te fluxes with varying substrate temperature. (a) substrate temperature, (b) chamber pressure (c) RHEED patterns during growth.

At t=0, the shutter is opened at a substrate temperature of 300 °C. The RHEED pattern does not qualitatively change. The spotty pattern typical for epitaxial growth appears only at temperatures close to 200 °C. Such a layer can be desorbed again during heating at a temperature of around 240 °C, which correlates with a corresponding peak in the chamber pressure. The diffuse streaky RHEED pattern typical of a stepped, two-dimensional surface is recovered at 300 °C. The

181

procedure can now be repeated, and a new layer can be regrown at substrate temperatures below 240 °C.
Despite the rough morphology and possible vacancies in the film, the layers show a good crystallinity and strongly dominant epitaxial orientation. An *in situ* x-ray plane scan normal to the surface along the [110] direction is shown in Fig. 5 for a film grown at 200 °C.

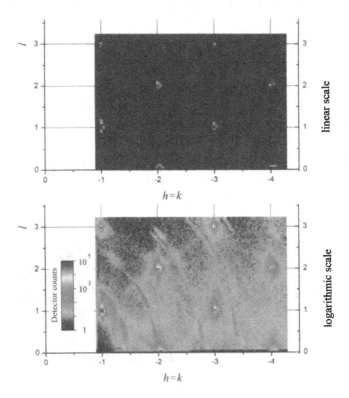

Figure 5. X-ray diffraction intensity map normal to the surface along the $h=k$ direction. The same data set shown in linear (top) and logarithmic (bottom) scale.

The epitaxial cubic orientation strongly dominates, both out-of-plane and in-plane (not shown, see [5]). Weak streaks due to grains in other orientations also originate from cubic crystallites. No peaks from the hexagonal phase are found.
Finally, we have switched the epitaxial layers to the amorphous phase using 60 ps pulses from a Nd-doped YAG laser. The samples were recrystallized by about 400 laser pulses at 20 Hz repetition rate at half the amorphization pulse energy. An optical microscope image of the switched regions is shown in Fig. 6. Regions #1 to #4 were repeatedly switched in both

directions, #1 four times, #2 two times and #3 and #4 three times. An untouched region is marked as #5. Too high intensity leads to a degradation as is visible in the center of region #1. The results demonstrate that the epitaxial films on GaSb(001) can be switched repeatedly as expected for GST.

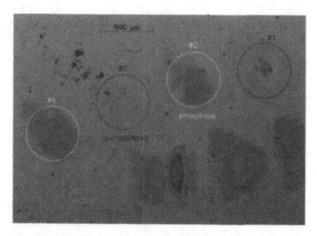

Figure 6. Optical microscope image of the epitaxial GST film shown in Fig. 1b after switching various regions by a pulsed laser.

ACKNOWLEDGMENTS

The authors would like to thank Steffen Behnke, Claudia Herrmann and Anne-Kathrin Bluhm. Support for this work was provided by the Deutsche Forschungsgemeinschaft (VR 1723/3-1) and the Japan Science and Technology Agency through an international research cooperation grant.

REFERENCES

1. G. I. Meijer, Science **319**, 1625 (2008).
2. A. V. Kolobov, P. Fons, A. I. Frenkel, A. L. Ankudinov, J. Tominaga and T. Uruga, Nat. Mat. 3, 703 (2004).
3. N. Yamada and T. Matsunaga, J. Appl. Phys. **88**, 7020 (2000).
4. S. Raoux, J. L. Jordan-Sweet, A. J. Kellock, J. Appl. Phys. **103**, 114310 (2008).
5. W. Braun, R. Shayduk, T. Flissikowski, M. Ramsteiner, H. T. Grahn, H. Riechert, P. Fons, A. Kolobov, Appl. Phys. Lett. **94**, 041902 (2009).
6. R. Shayduk, W. Braun, J. Cryst. Growth, in press.
7. B. Jenichen, W. Braun, V. M. Kaganer, A. G. Shtukenberg, L. Däweritz, C.-G. Schultz, K. H. Ploog, Rev. Sci. Instrum. **74**, 1267 (2003).

Mater. Res. Soc. Symp. Proc. Vol. 1160 © 2009 Materials Research Society   1160-H14-07

Crystallization Characteristics of Ge-Sb Phase Change Materials

Simone Raoux[1], Cyril Cabral[2], Jr., Lia Krusin-Elbaum[2], Jean L. Jordan-Sweet[2], Martin Salinga[3], Anita Madan[4] and Teresa L. Pinto[4]

[1]IBM/Macronix PCRAM Joint Project, Almaden Research Center, San Jose, California 95120, USA

[2]IBM T. J. Watson Research Center, Yorktown Heights, New York 10598, USA

[3] Physikalisches Institut (1A), RWTH University of Technology, 52056 Aachen, Germany

[4]IBM Hudson Valley Research Park, Hopewell Junction, New York 12533, USA

ABSTRACT

The crystallization behavior of Ge-Sb phase change materials with variable Ge:Sb ratio X between 0.079 and 4.3 was studied using time-resolved x-ray diffraction, differential scanning calorimetry, x-ray reflectivity, optical reflectivity and resistivity vs. temperature measurements. It was found that the crystallization temperature increases with Ge content from about 130 °C for X = 0.079 to about 450 °C for X = 4.3. For low X, Sb x-ray diffraction peaks occurred during a heating ramp at lower temperature than Ge diffraction peaks. For X = 1.44 and higher, Sb and Ge peaks occurred at the same temperature. Mass density change upon crystallization and optical and electrical contrast between the phases show a maximum for the eutectic alloy with X = 0.17. The large change in materials properties with composition allows tailoring of the crystallization properties for specific application requirements.

INTRODUCTION

Re-writable optical data storage based on phase change materials is a mature technology and much effort was spent in the design and optimization of the materials currently in use [1]. In the early stages of Phase Change Random Access Memory (PCRAM) technology the same materials that are used in optical storage (most prominently $Ge_2Sb_2Te_5$) were applied. However, PCRAM materials require a different set of materials properties compared to optical storage. In particular, the thermal stability and data retention are problematic for $Ge_2Sb_2Te_5$-based PCRAM devices. Failure times of PCRAM cells of 10 years at 85 °C have been reported [2], but these times become much shorter at higher temperature and are just a few days at 150 °C which is the operation temperature for automotive applications, and even just a few seconds at temperatures above 200 °C.

It has been reported recently that PCRAM devices with Ge-Sb phase change materials (with X = 0.17, the eutectic alloy) show very promising switching behavior [3]. The re-crystallization temperature of RESET devices was measured as a function of RESET conditions (the RESET operation is the melt-quench operation to switch the device to high resistance). It was observed that the crystallization temperature can be as high as 220 °C (depending on the RESET conditions) which is very promising in terms of data retention. In addition, very fast SET operations with pulses as short as 4 ns were reported (the SET operation is the re-crystallization to switch the device to low resistance). Here we have studied the crystallization properties of Ge-Sb alloys as a function of composition using time-resolved x-ray diffraction (XRD), differential scanning calorimetry (DSC), x-ray reflectivity (XRR), optical reflectivity (OR) and resistivity vs. temperature (R vs. T) measurements.

EXPERIMENTAL

Thin films of the phase change material Ge-Sb were deposited by co-sputtering from elemental targets with Ge:Sb ratios fractions ranging from 0.079 to 4.3 as determined by Rutherford Backscattering Spectrometry (RBS). The composition was varied by changing the relative powers of the Ge and Sb sputter sources. The substrates were Si wafers for XRD, XRR and OR measurements, photoresist-coated Si wafers for DSC measurements, and 1 μm SiO_2-coated Si wafers for R vs. T measurements. The material between the R vs. T contact pads was capped with 10 nm SiO_2 to prevent oxidation. Film thicknesses were between 50 – 100 nm, in a range where the crystallization temperature is independent of film thickness [4].

Time-resolved XRD measurements were performed at beamline X20C of the National Synchrotron Light Source at Brookhaven National Laboratory. The intensity of the diffracted XRD peaks was recorded during a heating ramp at a rate of 1 °C/s over 2θ range of 24 -39 °. The photon energy was 6.9 keV and photons were delivered by a high-throughput synthetic multilayer monochromator.

DSC analysis was performed on free-standing 100 nm Ge-Sb films using a DuPont DSC 2910 system. The films for DSC analysis were deposited on wafers which were coated with a 1 μm thick resist that was soft-baked. The photoresist was dissolved in acetone to lift off the Ge-Sb films which were then passed through a filter and dried overnight at room temperature. The Ge-Sb material was heated at a constant ramp rate, 10 °C/min, from 50 °C to 650 °C in a He ambient while heat flow was measured.

XRR was used to measure film thicknesses of as-deposited amorphous films and films heated above the crystallization temperature (as determined from the appearance of diffraction peaks in the time resolved XRD measurements). XRR data were collected using a high-resolution x-ray diffractometer with a monochromatic Cu K radiation source and simulated using commercially available software. A model consisting of a top oxide layer and the Ge-Sb layer on Si was used to fit the data. The thickness of the layers was obtained from the best fits. Mass density was then determined using the RBS results.

Optical reflectivity was measured using a custom-made instrument outfitted with a helium-neon laser at an incident angle close to normal (88°).

R vs. T was also measured using a custom-made setup consisting of two-probes contacting Al or TiN contact pads deposited on the films in a well-defined geometry (large pads with small gap). Resistivity was calculated from the sheet resistance, film thickness (measured by RBS) and electrode geometry. The measurements were performed in a nitrogen atmosphere.

186

RESULTS AND DISCUSSION

Figure 1 shows as an example the intensity of diffracted x-ray peaks during a heating ramp for a film with X = 0.37.

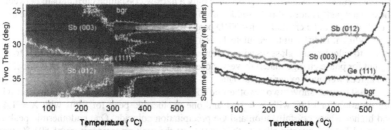

Figure 1. Left – intensity of diffracted x-ray peaks as a function of temperature over the 2θ range between 24 and 39 ° for a Ge-Sb film with a Ge:Sb ratio of 0.37. Right – Summed intensity integrated over angular ranges indicated on the left side corresponding to the Sb (003) and (012) peak and the Ge (111) peak normalized to the background (bgr).

It can be seen that the as-deposited film is amorphous and does not show any XRD peaks. At about 300 °C Sb peaks appear indicating the crystallization of the sample. At a higher temperature of about 370 °C the Ge (111) peak appears. This is correlated to Ge precipitation and elemental segregation [5, 6]. For films with low Ge content the Sb peaks appeared at a lower temperature than the Ge peaks, for films with X = 1.44 and higher Sb and Ge peaks occurred simultaneously. As shown in figure 2, the crystallization temperature T_x varied over a wide range from about 130 °C to about 450 °C with changing film composition and it increased with Ge content.

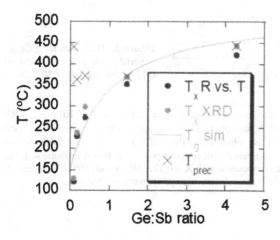

Figure 2. Measured crystallization temperature T_x from XRD and R vs. T measurements, simulated glass transition temperature T_g, and Ge precipitation temperature T_{prec} measured by XRD as a function of film composition.

This is in good agreement with literature [7, 8]. The Ge precipitation temperature (also shown in figure 2) did not vary much with film composition. In addition to T_x measured by XRD, T_x determined from R vs. T measurements (sudden drop in resistivity upon heating) is also plotted. The agreement is good, typically T_x determined from R vs. T measurements is slightly lower because only a narrow crystalline current path between the contact pads is needed for a sudden drop in resistivity while a substantial fraction of the film needs to be crystalline for XRD peaks to appear.

While it is difficult to calculate T_x as a function of film composition there is a model to calculate the glass transition temperature T_g [9, 10], which is a lower limit for T_x and for phase change materials very close to T_x [11]. This is also plotted in figure 2 and agrees well with experimental data.

DSC data show two exothermic peaks correlated to crystallization and Ge precipitation for films with $X < 1.44$ and one exothermic peak for films with $X = 1.44$ and higher where crystallization and Ge precipitation coincide. One endothermic peak is observed for the eutectic alloy with $X = 0.17$ at the melting temperature of 591 °C (see figure 3), while the alloys with lower Ge content have one endothermic peak at the melting temperature of Sb (around 630 °C) and alloys with higher Ge content show two endothermic peaks at the melting temperatures of the eutectic alloy and Sb. T_x and T_{prec} determined from DSC measurements are slightly lower than XRD and R vs. T measurements because the heating rate was lower.

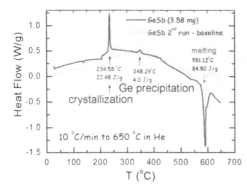

Figure 3. DSC scan for a Ge-Sb material with a Ge:Sb ratio of 0.17 heated at a rate of 10 °C/min.

The film mass density ρ was determined by a combination of RBS and XRR measurements for as-deposited, amorphous films (ρ_{amorph}) and films heated above their respective T_x but below T_{prec} for films where they differed (ρ_{cryst}). The mass density change ($\rho_{cryst} - \rho_{amorph})/\rho_{cryst}$ as a function of film composition is shown in figure 4. It has a maximum for the eutectic composition with $X = 0.17$. Alloys with high Ge content ($X = 1.44$ and higher) showed little mass density change upon crystallization.

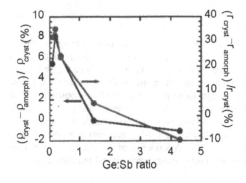

Figure 4. Change in mass density ρ and change in reflectivity r upon crystallization as a function of film composition.

The change in reflectivity r upon crystallization ($r_{cryst} - r_{amorph}$)/ r_{cryst} was also found to have a maximum of 36 % for the eutectic alloy (X = 0.17) and it is also plotted in figure 4. Low optical contrast was found for alloys with high Ge content.

Measurements of the resistivity as a function of temperature also revealed a maximum in the electrical contrast (ratio between the resistivity in the amorphous phase to the resistivity in the crystalline phase) as figure 5 shows.

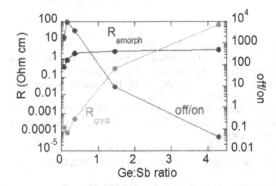

Figure 5. Resistivity in the amorphous and crystalline phases of Ge-Sb phase change materials and off/on ratio between the resistivities in the amorphous and crystalline phases as a function of Ge:Sb ratio.

The resistivities in both phases increase with increased Ge content, in particular the resistivity of the crystalline phase increases by more than five orders of magnitude. For the alloy with the highest Ge:Sb ratio measured here (X = 4.3) an inverse electrical contrast was measured. Even though there was a sudden drop in resistivity at the crystallization temperature the crystalline phase has a stronger dependence on the temperature than the amorphous phase so that after cooling back to room temperature the crystalline phase had a higher resistance than the original as–deposited amorphous phase. RBS performed in the area between the contact pads showed no change in material composition caused by annealing and crystallization, but additional change in resistance by reactions between the contact pads and the phase change material cannot be excluded.

SUMMARY

We have measured the crystallization behavior of Ge-Sb phase change materials with variable Ge:Sb ratio. Several parameters such as mass density change and optical and electrical contrast have a maximum for the eutectic composition with a Ge:Sb ratio of 0.17. The crystallization temperature increases monotonously with increased Ge fraction. Resistivities in both the amorphous and crystalline phase vary with composition and in particular the resistivity of the crystalline phase increases several orders of magnitude with increased Ge content leading to a reduced and even inversed electrical contrast.

This study reveals that it is possible to change Ge-Sb materials properties over a wide range by adjusting the composition. This dependence of the materials properties can be used to fine-tune crystallization behavior to meet technological requirements.

ACKNOWLEDGEMENTS

The authors thank Andrew Kellock for Rutherford Backscattering Spectrometry measurements. Research was carried out in part at the National Synchrotron Light Source, Brookhaven National Laboratory, which is supported by the U.S. Department of Energy, Division of Materials Sciences and Division of Chemical Sciences, under Contract No. DE-AC02-98CH10886.

REFERENCES

[1] M. Wuttig and N Yamada, Nature Mater. **6**, 824 (2007).
[2] B. Gleixner, A. Pirovano, J. Sarkar, F. Ottogalli, I. Tortorelli, M. Tosi and R. Bez, Proc. International Reliability Physics Symposium '07, p. 542 (2007)
[3] D. Krebs, S. Raoux, C. T. Rettner, R. M. Shelby, G. W. Burr and M. Wuttig, Mar. Res. Soc. Proc. Vol. 1072, paper 1072-G06-07 (2008)
[4] S. Raoux, J. L. Jordan-Sweet, and A. Kellock, J. Appl. Phys. **103**, 114310 (2008).
[5] C. Cabral, Jr., L. Krusin-Elbaum, J. Bruley, S. Raoux, V. Deline, A. Madan, and T. Pinto, Appl. Phys Lett. **93**, 071906 (2008)
[6] S. Raoux, C. Cabral, Jr., L. Krusin-Elbaum, J. L. Jordan-Sweet, K. Virwani, M. Hitzbleck, M. Salinga, A. Madan, and T. L. Pinto, J. Appl. Phys. **105**, 064918 (2009)
[7] T. Okabe, S. Endo, and S. Saito, J. Non-Cryst. Solids **117/118**, 222 (1990).
[8] J. M. Del Pozo, M. P. Herrero and L. Diaz, J. Non-Cryst. Solids **185**, 183 (1995)
[9] M. H. R. Lankhorst, J. Non-Crystall. Solids **297**, 210 (2002).
[10] S. Raoux, M. Salinga, J. L. Jordan-Sweet, and A. Kellock, J. Appl. Phys. **101**, 044909 (2007)
[11] J. Kalb, M. Wuttig, and F. Spaepen, J. Mater. Res. **22**, 748 (2007)

Mater. Res. Soc. Symp. Proc. Vol. 1160 © 2009 Materials Research Society   1160-H14-11

Conformal MOCVD Deposition of GeSbTe in High Aspect Ratio Via Structures for Phase Change Memory Applications

J. F. Zheng[1], P. Chen[1], W. Hunks[1], M. Stender[1], C. Xu[1], W. Li[1], J. Roeder[1]
S. Kamepalli[2], C. Schell[2], J. Reed[2], J. Ricker[2], R. Sandoval[2], J. Fournier[2],
W. Czubatyj[2], G. Wicker[2], C. Dennison[2], S. Hudgens[2], T. Lowrey[2]

[1]ATMI Inc., 7 Commerce Drive, Danbury, CT 06810
[2]Ovonyx Technologies, Inc., 2956 Waterview Drive, Rochester Hills, MI 48309

ABSTRACT

We have demonstrated conformal deposition of amorphous GeSbTe films in high aspect ratio structures by MOCVD. SEM analysis showed the as-deposited GeSbTe films had smooth morphologies and were well controlled for void free amorphous conformal deposition. GeSbTe films adhere well to SiO_2, TiN, and TiAlN. The morphology and adhesion are stable in 420°C post process. By annealing at 365°C, amorphous GeSbTe films converted into crystalline GeSbTe with polycrystalline grain sizes of 5nm. Film resistivity in the crystalline phase ranged from 0.001 to 0.1 Ω-cm, suitable for device applications. Phase change devices fabricated with confined via structures filled with MOCVD GeSbTe showed cycle endurances up to 1×10^{10} with a dynamic set/rest resistance of two orders of magnitude.

INTRODUCTION

Since its invention by Ovshinsky, PCM (Phase Change Memory) based on GST alloys (e.g. $Ge_2Sb_2Te_5$) has become one of the most promising candidates for coming generation nonvolatile memory applications due to its scalability when compared with current charge based memory [1]. Certain DRAM applications are feasible due to the demonstrated high intrinsic cycle endurance, speed, and path for reduction of reset current [2-4]. Planar PCM devices made by Physical Vapor Deposition (PVD) of GST alloys demonstrated good characteristics of phase change behavior, with performance surpassing the requirements for flash memory [5]. PCM in a confined cell structure with high aspect ratio (> 3:1) and sub-100nm diameters have been filled with Chemical Vapor Deposition (CVD) deposited GST and have substantial advantages over cell designs with planar structures made by PVD deposited GST alloys [3,4]. In a CVD GST based confined PCM cell of dash shape with a cross-section of 7.5nm x 60nm and depth of 30nm, a low reset current of 160uA, high cycling endurance of 1×10^{10}, and fast set speed of 50ns have been demonstrated [4]. The key factor is the conformal filling of very high aspect ratio confined structure by CVD or Atomic Layer Deposition (ALD) [6, 7]. Successful adoption of CVD or ALD deposited GST materials for PCM applications will require suitable material properties in addition to achieving a conformal fill of the confined structures. These requirements include: void free deposition, good adhesion to SiO_2 and top or bottom metal electrode, material properties stable at 400°C to survive subsequent device processing steps, low resistivity in the crystalline state, and possess a wide resistivity range between the amorphous and crystalline phases, which is desired for Multi-Level-Cell (MLC) applications.

In this paper, we examine the properties of MOCVD deposited GST films with respect to the above attributes for application in PCM. Test PCM devices fabricated with confined structures on integrated devices demonstrated high cycle endurance and a wide dynamic range of set/reset resistance.

EXPERIMENT

GST films were deposited by MOCVD on substrates with SiO_2, Si_3N_4, TiN or TiAlN coated surfaces on both planar and various trench and via structures. Compositions were determined by Wavelength Dispersive X-Ray Fluorescence (XRF), where the XRF was calibrated using sputtered PVD GST films of known thickness and composition. Film morphology and degree of conformality was studied by cross-sectional SEM or STEM. Amorphous to crystalline transformations were performed by annealing in a N_2 purged tube furnace and measured by X-Ray Diffraction (XRD). Film resistivity before and after annealing was calculated from the film thickness and the sheet resistance measured using a four point probe.

Functioning PCM devices were made by filling 70nm diameter and 200nm via structure with MOCVD deposited GST films. Following the GST conformal fill, a TiN top electrode was deposited by PVD sputtering; the bottom electrode was TiAlN. Subsequently, the TiN and GST layers were patterned and etched to form isolated top electrodes for single PCM cell test devices. An Al layer was subsequently deposited and patterned such that it completely overlapped the underlying TiN/GST electrode thus functioning as both a probe pad and encapsulation layer for the GST material.

DISCUSSION

GST Film Morphology

A cross-sectional STEM image of a smooth and void free as-deposited MOCVD GST film on SiO_2 and covered by a sputtered TiN top layer on top is shown in figure 1a. A cross-sectional SEM image of the same sample following a 420°C anneal for 30min in a N_2 purged protecting environment is shown in figure 1b. The SEM images show that the GST films were also smooth and void free after annealing. The GST/SiO_2 and GST/TiN interfaces are intact and free from delamination after the 420°C anneal, similar to that of the as-deposited film.

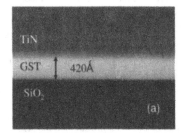 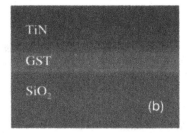

Figure 1. (a) Cross-Sectional STEM image of GST film as-deposited on a SiO_2 substrate and covered by TiN film. (b) Same film after annealing in a furnace at 420°C for 30min.

Film Crystallinity

As-deposited MOCVD GST films are preferably amorphous in order to control the conformal deposition and morphology. X-Ray diffraction spectra of a typical MOCVD deposited amorphous GST film is shown in figure 2a. After the GST film was annealed in a tube furnace under nitrogen for 45min at 365°C, the X-Ray diffraction spectrum clearly shows crystalline GST fcc structure peaks, figure 2b.

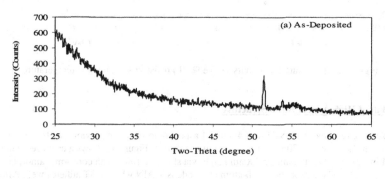

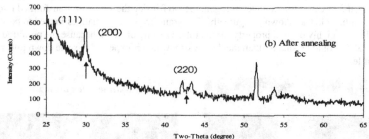

Film Resistivity

As-deposited MOCVD GST films are highly resistive. For films of 500Å or less, sheet resistance is typically outside the measurement range of 20MΩ/□, equivalent to about 100 Ω-cm resistivity. After the GST film is annealed in a tube furnace under a 200sccm nitrogen purge for 45min at 365°C, the sheet resistivity is drastically reduced to the desired 0.1 Ω-cm or less. Resistivity data after annealing plotted against a selected section of Te compositions ranging from 39% to 46% is depicted in figure 3a. The wide range of resistivity values between 0.001 to 0.1 Ω-cm is due to variation in the Ge and Sb compositions. Tuning of the Ge and Sb composition for proper resistivity value as required by device design is achievable by MOCVD. Resistivity vs. inverse temperature in figure 3b shows the activation energy for resistivity is very small, about 10meV- consistent with reported values for annealed PVD GST films [8].

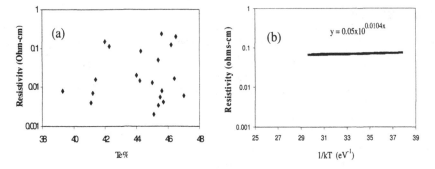

Figure 3. (a) measured resistivity vs. Te %. (b) resistivity vs. inverse temperature.

GST Fill of High Aspect Ratio Structure

We have reported the GST fill of 40nm 5:1 aspect ratio structure previously [9]. Here we are focusing on GST fill of a 70nm 3:1 aspect ratio test device. Figure 4a shows a cross-sectional STEM image of a 70nm diameter 200nm height via structure filled with conformal amorphous GST using a MOCVD process. The bottom electrode is TiAlN where GST adheres well. After annealing at 375°C for 30min under nitrogen, the GST inside the via phase transformed into poly-crystalline GST as shown in figure 4b. The grain size is very small, estimated to be about 5nm. This is a highly desired property for the future scaling of PCM, because the preferred grain size for GST should be smaller than the device size, which is expected to scale down toward 10nm or less.

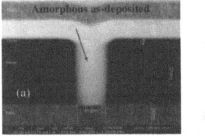
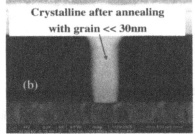

Figure 4. (a) GST filled structure as-deposited. (b) GST filled structure after annealing.

Figures 5a and 5b show normalized compositions measured by EDS with quite uniform distribution of Ge and Te across the via from position 1 to 5, which is from the top to the bottom of the via structure. Compositions at position 6 and 7, outside and away from the via are also similar to that from 1 to 5, showing excellent composition distribution during the conformal deposition of the film. The measured Sb distribution has larger variation, possibly due to measurement inaccuracy for Sb.

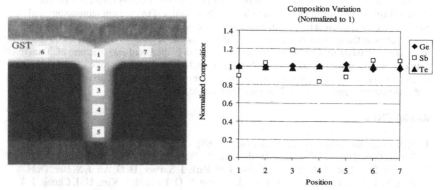

Figure 5. (a) EDS measurement locations. (b) Normalized composition variations.

Device Performance

Figure 6a illustrates the PCM test device structure with confined via filled with the MOCVD GST. Figure 6b and 6c are the I-V and R-I curves of a selected PCM device. The device has a reset current of 1mA , Vt of 1.36V. Rreset/Rset ratio> 100, and parameter cycle endurance of 1×10^{10} as indicated in Figure 6d.

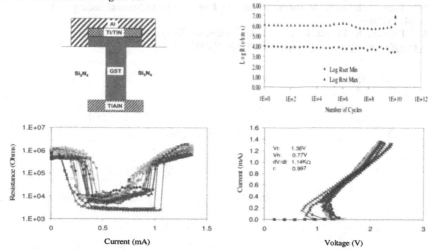

Figure 6. (a) PCM device structure. (b). I-V curves of PCM device. (c). R-I curves of PCM device. (d) Cycle endurance of program parameters.

195

CONCLUSION

Smooth and void-free GST films have been deposited successfully by MOCVD. The MOCVD films exhibit good adhesion to SiO_2, TiN as well as TiAlN. Film morphology and adhesion are compatible with post deposition processing requirement of 400°C. Annealing at 365°C converts the as-deposited amorphous GST films into crystalline GST films with resistivity in the range of 0.001 to 0.1 Ω-cm. Test devices with GST filled confined structure of 70nm in diameter and 3:1 aspect ratio structures have good performance with the following characteristics: 1mA reset current, 1.37 Vt, parameter cycle endurance of up to 1×10^{10} cycles, and resistance dynamic range of 2 orders of magnitude.

REFERENCES

1. S. R. Ovshinsky, *Phys. Rev. Lett.* **21**, 1450 (1968).
2. S. Lai, Tech. Dig. Int. Electronic Devices Meet. 2003, 255.
3. J. I. Lee, S. L. Cho, Y. L. Park, B. J. Bae, J. H. Park, J. S. Park, H. G. An, J. S. Bae, D. H. Ahn, Y. T. Kim, H. Horrii, S. A. Song, J. C. Shin, S. O. Park, H. S. Kim, U.-I. Chung, J. T. Moon, and B, I. Ryu, Symp. of VLSI Tech. Dig., 2007, 102
4. D. H. Im, J. I. Lee, S. L. Cho, H. G. An, D. H. Kim, I. S. Kim, H. Park, D. H. Ahn, H. Horii, S. O. Park, U-In Chung, and J. T. Moon, Dig. Int. Electronic Devices Meet. 2008, 211
5. A. Pirovano F. Pellizer, I. Tortorelli, R. Harrigan, M. Magistretti, P.Petruzza, E. Varesi, D. Erbetta, T. Marangon, F. Bedeschi, R. Fackenthal, G. Atwood, and R. Bez, Solid-State Electronics, **52**, 1467 (2008).
6. B.J. Choi, S. Choi, Y. C. Shin, C. S. Hwang, J. W. Lee, J. Jeong, Y. J. Kim, S.-Y. Hwang, And S. K. Hong, J. Electrochem. Soc. **154**, H318 (2007).
7. V. Pore, T. Hatanpaa, M. Ritala, M. Leskela, J. of American Chemical Society, **131**, 3478 (2009)
8. T. Kato and K. Tanaka, Japanese, J. Appl. Phys. **44**, 7340(2005).
9. http://www.semiconductor.net/article/CA6621100.html:

Printed in the United States
By Bookmasters